Solar Pumping for Water Supply

Praise for this book

'This book is a truly fantastic "one stop shop" for all solar water pumping needs; it covers examples from humanitarian and developments across the entire globe and deals with both simple and easy-to-follow rules of thumb and also extremely detailed design parameters. There is no other book on solar water pumping with the breadth and depth that this one covers in such a practical and down-to-earth way. It's an essential reading and reference book for anybody designing and installing solar water systems.'

Andy Bastable, Head of Water & Sanitation, Oxfam

'*Solar Pumping for Water Supply* is an excellent book that brings together a perfect merger of the theory and practice of the subject matter. It provides a clear road map from the project conceptualisation, its design, implementation including the social impact of such projects. With engineering formulae and photographic illustrations it goes to provide excellent examples of how to and how not to do Solar pumping water supplies, with cases drawn from across Africa and Asia. I highly recommend the book for practitioners and learners of water supply and solar renewable energy as it provides the fusion of the two disciplines to deliver the scare water resources in the most economical manner.'

Dr MAS Waweru, Managing Director Davis & Shirtliff Ltd

'This is a good reference book to be used by anybody – not just technicians – keen on knowing more about all aspects of solar PV pumping in emergency and developments projects, going from technology, design, installation, maintenance and financing of solar pumping systems. The book also provides interesting and useful case studies in the annexes.'

Jean-Paul Louineau, Director, Alliance Soleil

'A great book that reflects the experience of the authors in the energy and humanitarian sector and that will surely be very well received by those who have to work in the implementation of photovoltaic systems for water pumping.'

Dr. Salvador Seguí-Chilet, Univ. Politécnica de Valencia

'This is a very timely and comprehensive guide to support the design, siting, procurement, installation, commissioning, operation and maintenance and monitoring of solar powered water systems. This guide will help to reduce the technical issues arising from inadequately designed solar powered water systems which have impeded the full utilisation of solar powered water systems to ensure the quality, equity and sustainability of safe water services.'

Silvia Gaya, Senior Advisor Water and Environment,
WASH Programme Division, UNICEF HQ

Solar Pumping for Water Supply
Harnessing solar power in humanitarian and development contexts

Asenath W. Kiprono and Alberto Ibáñez Llario

Practical Action Publishing Ltd
27a, Albert Street, Rugby,
Warwickshire, CV21 2SG, UK
www.practicalactionpublishing.com

© Asenath W. Kiprono and Alberto Ibáñez Llario, 2020

This open access article is distributed under a Creative Commons Attribution Non-commercial No-derivatives CC BY-NC-ND license. This allows the reader to copy and redistribute the material; but appropriate credit must be given, the material must not be used for commercial purposes, and if the material is transformed or built upon the modified material may not be distributed. For further information see https://creativecommons.org/licenses/by-nc-nd/4.0/legalcode

Product or corporate names may be trademarks or registered trademarks, and are used only for identification and explanation without intent to infringe.

A catalogue record for this book is available from the British Library.

A catalogue record for this book has been requested from the Library of Congress.

ISBN 978-1-78853-036-1 Paperback
ISBN 978-1-78853-035-4 Hardback
ISBN 978-1-78044-781-0 Library PDF
ISBN 978-1-78044-782-7 Epub

Citation: Kiprono, A W., Llario, A I., (2020) *Solar Pumping for Water Supply: Harnessing solar power in humanitarian and development*, Rugby, UK: Practical Action Publishing <http://dx.doi.org/10.3362/9781780447810>.

Since 1974, Practical Action Publishing has published and disseminated books and information in support of international development work throughout the world. Practical Action Publishing is a trading name of Practical Action Publishing Ltd (Company Reg. No. 1159018), the wholly owned publishing company of Practical Action. Practical Action Publishing trades only in support of its parent charity objectives and any profits are covenanted back to Practical Action (Charity Reg. No. 247257, Group VAT Registration No. 880 9924 76).

The views and opinions in this publication are those of the author and do not represent those of Practical Action Publishing Ltd or its parent charity Practical Action. Reasonable efforts have been made to publish reliable data and information, but the authors and publisher cannot assume responsibility for the validity of all materials or for the consequences of their use.

Cover photos: Top photo shows IOM water scheme at Kutupalong Balukhali Expansion Site refugee camp, courtesy of IOM Bangladesh. Bottom photo shows villagers in Darfur, Sudan with a new solar water pump – part of an Integrated Water Resources Management (IWRM) system that came out of a collaboration between Practical Action, local governments, technical departments and the communities affected by drought, courtesy of Practical Action.

Cover design by RCO.design
Typeset by vPrompt eServices, India

Contents

Boxes, figures, and tables	ix
Acronyms	xv
Preface	xvii

1.	Solar photovoltaic solutions for water pumping	1
	1.1 Solar PV water pumping in humanitarian and development contexts	1
	1.2 Factors influencing the renewed interest in solar PV water pumping	3
	1.3 Guidance note on the use of solar pumping	5
2	Definitions and principles of solar energy production	9
	2.1 The solar resource	9
	2.2 Sun and water: the perfect relationship	10
	2.3 Solar radiation	11
	2.4 Solar photovoltaic	12
	2.5 Solar irradiance	14
	2.6 Solar insolation	18
	2.7 Standard test conditions	18
	2.8 Solar resource maps for peak sun hours	21
	2.9 Basic DC electric concepts	23
	2.10 Solar module I-V curve and maximum power point	25
3	Solar-powered water system configurations and components	27
	3.1 SPWS concept and revolution	27
	3.2 SPWS configurations	28
	3.3 SPWS components	32
	3.4 SPWS balance-of-system components	42
	3.5 SPWS equipment manufacturers	46
	3.6 Importance of quality considerations in SPWSs	47
4	Energy losses in solar photovoltaic energy production	49
	4.1 Calculating energy losses	49
	4.2 Cell temperature energy losses	51
	4.3 Wiring energy losses	54
	4.4 Sun irradiance energy losses	56
	4.5 PV module energy losses	61
	4.6 Module mounting energy losses	63
	4.7 Power converters and the balance-of-system energy losses	64
	4.8 Estimation of the energy yield	66
5	Design of a solar-powered water scheme	67
	5.1 Solar pump design	67
	5.2 Important design concepts and considerations	68
	5.3 Steps to design a solar-powered water scheme	69

6	Electrical and mechanical installation of solar-powered water systems	87
	6.1 Pumping system installation	87
	6.2 Installation sequence and process	89
	6.3 Earthing, lightning protection, and surge protection	106
	6.4 Electrical safety	108
7	Specific considerations and limitations for solar-powered water pumping	109
	7.1 Chlorination	109
	7.2 Full-tank detection at long distances	111
	7.3 Solar tracking	112
	7.4 Emergency solar pumping kits	112
	7.5 Power range for pump motors and inverters	113
	7.6 Vandalism and theft	114
	7.7 Overpumping of aquifers	117
	7.8 Hot climate zones and hot water pumping	118
	7.9 Frequently asked questions	120
8	Solar-powered water pumping for agriculture	123
	8.1 Water for irrigated agriculture	123
	8.2 The influence of pressure on energy requirements in irrigation	123
	8.3 Greenhouse gas emissions from agriculture and climate-change adaptation	125
	8.4 Financing solar-powered irrigation systems	126
	8.5 Financing instruments to develop solar irrigation	127
	8.6 The risks and challenges of solar irrigation	128
	8.7 Recommendations for solar irrigation challenges	129
9	Economic analysis: life-cycle cost of different pumping technologies	133
	9.1 The importance of economic considerations	133
	9.2 Life-cycle cost analysis	134
	9.3 Life-cycle costing for water pumping	135
	9.4 Comparing LCCA of solar and generator systems	138
	9.5 Cost of ownership	144
10	Calls for proposal and bidding	147
	10.1 Selection criteria for solar pumping products and services	147
	10.2 Desired features of key components	147
	10.3 Supplier selection	150
	10.4 Bidding process	151
	10.5 Bidding template: technical terms of reference	152
	10.6 Quality of solar modules	156
	10.7 Practical aspects of equipment and supplier selection	160
11	Testing and commissioning, operation and maintenance	163
	11.1 Testing and commissioning	163
	11.2 Operation and maintenance of equipment	167

12 Warranties, social models for management, and monitoring	179
12.1 Warranties	179
12.2 Social models for management	182
12.3 Monitoring	183
Annex A Pump and generator design basics	187
Annex B Manual calculation of solar system	195
Annex C Example of calculation of losses due to non-optimum tilt angle of PV modules	211
Annex D Cable sizing	215
Annex E Product warranty card sample	219
Annex F Routine inspection and maintenance activity sheets	221
Annex G Preventive maintenance plan	225
Annex H General troubleshooting for SPWSs	229
Annex I Financing instruments for solar-powered irrigation systems	233
Annex J Physical control installation and maintenance checklists	237
Annex K Daily photovoltaic module and pump operation/monitoring format	241
Glossary	243
References	249

Boxes, figures, and tables

Boxes

Box 2.1 Solar PV components	13
Box 4.1 Minimizing losses due to temperature	55
Box 4.2 Minimizing wiring losses	57
Box 4.3 Minimizing losses due to soiling	59
Box 4.4 Minimizing losses due to shading	59
Box 4.5 Minimizing losses due to incorrect azimuth and tilt angle	61
Box 4.6 Minimizing mismatching losses	62
Box 4.7 Minimizing ageing losses	63
Box 4.8 Minimizing losses due to availability	65
Box 10.1 Dimensions of product quality	148
Box 10.2 Criteria for supplier selection	150
Box 10.3 A reference guide for the SPWS procurement process	152
Box 10.4 Checking module quality	160
Box 11.1 When to call a technician	172
Box 11.2 Example of essential provisions in a maintenance service agreement	175
Box 12.1 Commonalities between successful community-managed water schemes in Kenya	183
Box 12.2 For local community projects, should solar water be free?	183

Figures

Figure 1.1 Global map of appropriate locations for solar applications	2
Figure 1.2 Evolution of price of PV cells, 1977–2015	4

x   SOLAR PUMPING FOR WATER SUPPLY

Figure 2.1   Sun and water: a perfect relationship   10
Figure 2.2   Global Horizontal Irradiation   11
Figure 2.3   Conversion of sunlight to electricity   12
Figure 2.4   The 'PV effect' process   13
Figure 2.5   Illustration of PV cell, module, array, and generator   14
Figure 2.6   A typical solar-powered water system   15
Figure 2.7   Weekly irradiation in Valencia in the month of July   16
Figure 2.8   Time-based solar irradiance in Valencia during summer
             for two consecutive days   17
Figure 2.9   Graph of daily insolation   18
Figure 2.10 Solar irradiance during two days of July and
             corresponding insolation   19
Figure 2.11 Air mass for different sun positions   20
Figure 2.12 PSH map for Jordan   22
Figure 2.13 Typical I-V and power curves for a crystalline PV
             module operating at STC   25
Figure 2.14 I-V curve under varying irradiance and temperature   26
Figure 3.1   Typical flow output profile of a solar-powered pump   29
Figure 3.2   Stand-alone solar installation   30
Figure 3.3   Hybrid installation   30
Figure 3.4   Stand-alone surface Installation   31
Figure 3.5   Illustration of a solar module construction   38
Figure 3.6   Different types of solar modules (clockwise from
             top left: mono-Si, poly-Si, a-Si and TFPV)   40
Figure 3.7   Parts of a cable   44
Figure 3.8   The components of a solar water pumping system   46
Figure 3.9   Worn-out conduit due to high temperatures in Iraq   48
Figure 4.1   Temperature difference for a c-Si plant with a fixed-tilt
             angle of 30° in Valencia, Spain   52
Figure 4.2   Effect of sun irradiance in efficiency of c-Si PV modules   53

BOXES, FIGURES, AND TABLES   xi

Figure 4.3 Spacers between modules to increase cooling   55

Figure 4.4 Sand on PV modules limiting power output   58

Figure 4.5 Shading of PV modules limiting power output   58

Figure 4.6 Example of warranty given by Trina Solar for Tallmax
            monocrystalline modules   63

Figure 4.7 Bypass diode activation in shadowed PV modules   64

Figure 5.1 Series connection   75

Figure 5.2 Parallel connection   75

Figure 5.3 Combination of series and parallel connections   76

Figure 5.4 Module array layout   77

Figure 5.5 Sun's position in the sky with respect to the earth's surface   79

Figure 5.6 Effect of tilt on solar energy capture   80

Figure 5.7 PV modules in Sheikhan, Iraq, facing south at 36° tilt   81

Figure 5.8 Inverter box located under modules in Somali
            Region, Ethiopia   82

Figure 5.9 Hybrid solar-generator scheme at Adjumani refugee
            settlement in Uganda   85

Figure 6.1 Sample nameplates for (left) PV module, and (right)
            solar controller   88

Figure 6.2 Pump installation using a hydraulic winch   91

Figure 6.3 Horizontal surface pumps installed on a concrete plinth
            in Itang Water Supply, Ethiopia   93

Figure 6.4 Installed controls (left to right): AC changeover switch,
            inverter, surge protector, PV disconnection switch   94

Figure 6.5 Inverter installed in a lockable meshed enclosure under
            the PV array at Turkana, Kenya   95

Figure 6.6 Examples of poor cable management in South Sudan   97

Figure 6.7 Examples of good cable management in South Sudan and
            Tanzania respectively   97

Figure 6.8 Ground mount structure in Kawrgosk refugee camp, Iraq   98

Figure 6.9 Pole mounted PV modules in Bidibidi refugee settlement,
            Uganda   98

xii SOLAR PUMPING FOR WATER SUPPLY

Figure 6.10 Roof mount on a flat tank roof in Gaza   99
Figure 6.11 Combined roof and pole mount installation at
             Rohingya Mega camp (Cover photo)   100
Figure 6.12 Example of ground screw structure   100
Figure 6.13 Tank-mounted structure in an IDP camp near
             Maiduguri, Nigeria   102
Figure 6.14 Failed pole mount in Yida, South Sudan, modules
             blown off in Fafen, Ethiopia (right)   103
Figure 6.15 A tight connection of solar module MC3 quick connectors   103
Figure 6.16 Leap-frog connection of modules   104
Figure 6.17 Example of untidy and tidy module cables   105
Figure 6.18 Walkway on a roof mount at Strathmore University, Kenya   105
Figure 6.19 Walkway on a ground mount in Koboko, Uganda   106
Figure 6.20 Lorentz surge protection device   107
Figure 7.1   Valve-regulated bypass chlorinator   111
Figure 7.2   Solar trackers: double-axis tracking (left) and single-axis
             tracking (right)   112
Figure 7.3   Fenced SPWS with a solar light for security in Turkana, Kenya   115
Figure 7.4   Extreme module security measures in Burao, Somalia   116
Figure 7.5   Awareness raising of the value of the SPWS to avoid theft
             and vandalism   117
Figure 8.1   Yield response of crops to water availability in China   124
Figure 8.2   Approximate shares of greenhouse gas emissions (in CO_2
             equivalent) emitted by the global agri-food sector in 2010   125
Figure 9.1   Steps for technical design and economic appraisal   137
Figure 9.2   Comparison over time of costs for solar vs generator   141
Figure 9.3   LCCA of stand-alone generator vs hybrid vs solar at
             Nyarugusu, Tanzania   144
Figure 9.4   Business models for financing solar-powered water systems   144
Figure 10.1 Quality of solar panels matters   148
Figure 10.2 Example of a solar-powered water system layout
             provided to bidders   153

Figure 10.3	Cheap will become expensive with time	161
Figure 11.1	Duration of different phases in the life of a solar pumping scheme	168
Figure 11.2	Routine maintenance activities to ensure proper daily functionality	171
Figure 12.1	Lorentz communication system for monitoring from phone via Bluetooth or from computer via Internet	184
Figure 12.2	Fluke clamp meter	186
Figure A1	Layout of total dynamic head calculation	188
Figure A2	Layout of worked example	191
Figure B1	Number of modules arranged in series and parallel	197
Figure B2	Grundfos sizing result	199
Figure B3	Lorentz sizing result	200
Figure B4	Grundfos SP 9 head-flow curve	201
Figure B5	Grundfos SP 9 pump data sheet	202
Figure B6	Grundfos SP 9 power-flow curves	203
Figure B7	Grundfos motor data sheet	204
Figure B8	Solar resource map for Kenya	205
Figure D1	Table of cable current capacity and voltage drop	216
Figure D2	Table of cable sizes	217

Tables

Table 1.1	Advantages and disadvantages of solar PV water pumping schemes	3
Table 2.1	Analogy of electricity and water	24
Table 3.1	List of solar-powered water systems found in the field	28
Table 3.2	Example of module characteristics	40
Table 4.1	Table of estimated losses as a percentage of total energy produced	50
Table 4.2	NOCT, g, and efficiency factors for different PV module technologies	54

Table 4.3 Variation of resistivity and conductivity with temperature	56
Table 5.1 Data needed to design a solar-powered water scheme	68
Table 5.2 Expected lifetime of solar components	70
Table 5.3 Example of worst-month calculation where constant water required: Yumbe, Uganda (3N, 31E)	72
Table 5.4 Example of worst-month calculation where variable water required: Yumbe, Uganda (3N, 31E)	72
Table 5.5 Quick guidance for water tank capacity	83
Table 7.1 Summary of Global Solar and Water Initiative specifications for small, medium-sized, and large solar pumping kits for rapid deployment	113
Table 9.1 Data required for life-cycle analysis	136
Table 9.2 Estimated cost for maintenance of a good-quality diesel generator	137
Table 9.3 Estimated generator fuel consumption at different loads	138
Table 9.4 LCCA for the given water scheme with a diesel generator	139
Table 9.5 Capital cost of the main components of the given solar PV pumping system	140
Table 9.6 LCCA for the given water scheme with a solar PV pumping system	140
Table 9.7 Cost comparison between existing generator stand-alone systems in South Sudan and equivalent solar or hybrid systems	143
Table 10.1 Minimum recommended characteristics of solar components	149
Table 11.1 Inspections and functionality tests for solar-powered water systems	164
Table 11.2 System performance tests for a solar-powered water system	165
Table 11.3 SPWS testing report template	166
Table 11.4 Documentation set accompanying the solar water-pumping scheme	169
Table 11.5 Records of service and maintenance activities at site level	170
Table 11.6 Routine maintenance activities for solar-powered water systems	171

Acronyms

BEP	best efficiency point
BoS	balance of system
FAO	Food and Agriculture Organization
FiT	feed-in tariff
GLOWSI	Global Solar and Water Initiative
GHG	greenhouse gas
IDP	internally displaced person
LCCA	life-cycle cost analysis
LCB	linear current booster
MPP	maximum power point
MPPT	maximum power point tracking
NOTC	nominal operating temperature of a cell
O&M	operation and maintenance
PPA	power purchase agreement
PSH	peak sun hours
PW	present worth
SPIS	solar-powered irrigation system
SPWS	solar-powered water system
STC	standard test conditions
SPD	surge protection devices
TDH	total dynamic head
TFPV	thin-film photovoltaic cells

Preface

While solar water pumping has been in operation since the 1970s, it is only in last few years that it has expanded globally, offering more robust, larger and efficient solutions for water supply projects. Tens of thousands of solar pumping schemes have been installed in the last decade, in both rural communities and in large urban settings, as well as in camps for internally displaced people and refugees; in emergency and post-emergency contexts and also in more developmental situations.

This book is based on five years' work developed within the framework of the Global Solar and Water Initiative. This Initiative is designed to mainstream quality solar pumping solutions in low- and medium-income countries, working hand in hand with governments, the private sector, manufacturers, academic institutions, NGOs and United Nations agencies.

In writing this book we have examined current technology, best practices, product quality and availability; we analysed costs and compared them with other available technologies; and we reviewed different operation and maintenance models. In the process, over a hundred IDP and refugee camps and communities were visited in 12 different countries, and hundreds of engineers were trained either in onsite events or via dedicated online training courses.

This book is the result of all this work. We have tried to explain in simple and clear language not only the theoretical knowledge needed to understand the technology, but also the practicalities and lessons learnt through all the visits, meetings and interviews carried out since 2016 relating to solar water pumping in humanitarian and development contexts. All the material developed within the Global Solar and Water Initiative, together with news about trainings helpline and other resources can be found at www.thesolarhub.org.

We extend our thanks to Albert Reichert and Jonathan Hamrell, from the Bureau for Humanitarian Assistance in USAID, who provided the support and the funds to write and publish this book.

We want to thank the hundreds of people we have met in the field in recent years for their work, interest, time and curiosity about solar water pumping solutions.

We are grateful to the publishers, Practical Action Publishing, who have been a source of guidance during the whole process of writing and publishing this book.

Special acknowledgement is due to Professor Salvador Seguí Chilet (Polytechnic University of Valencia, Spain) who authored chapter 4, and Florent Eveillé (FAO) who authored chapter 8. We thank the following people who provided technical review and gave invaluable feedback: Brian McSorley (Oxfam), Kai Rainecke (Lorentz), Andrew Armstrong and Jeff Zapor (Water Mission), Antonio Torres (IOM) and Professor Ellen Milnes (University of Neuchatel, Switzerland). Finally, we thank Jerome Burlot, who provided great support in starting the Global Solar and Water Initiative back in 2016, together with his colleagues Daniel Clauss and Denis Heidebroek (ECHO).

Our deepest gratitude goes to our families, Ezekiel, Kuan-Yun and Arnaud, for their unwavering support and sacrifice in allowing us to work long hours and to depart on long work trips, while they often shouldered greater responsibilities. To Lisa, Joshua, Nina and Noa – thank you for allowing us time away from you to work on this book.

CHAPTER 1
Solar photovoltaic solutions for water pumping

Following a worldwide energy transition to renewable solutions, humanitarian and development actors are increasingly using solar photovoltaic technology in their water supply projects. A number of factors, including reduced costs, reliable technology, a booming private sector, high solar radiation in vast areas of Africa and Asia, and environmental concerns, among others, have been pivotal to bringing about this renewed interest in solar PV solutions in the relief sector. Despite its numerous advantages, solar PV pumping is not a panacea and careful contextual analysis beyond technical considerations should be carried out before its adoption.

Keywords: energy transition, humanitarian energy, solar pumping guidance, solar pumping advantages, solar cost reduction

1.1 Solar PV water pumping in humanitarian and development contexts

An energy transition is well on the way worldwide, with an exponential expansion in the use of renewable energy solutions in both developed and developing countries. This expansion is creating local value and jobs, mitigating climate change, and creating stronger community resilience (REN21, 2019).

The supply of constant and reliable energy is of paramount importance to ensure the provision of basic services, such as water. In a large number of water supply projects, water is pumped from boreholes or surface water bodies to elevated storage tanks and fed into distribution systems using the force of gravity. Especially in developing countries, the energy required for pumping water is dependent on generators and fuel and/or electrical grids that are unreliable or faulty. Due to deep groundwater tables, distant water points, and/or the requirement for large quantities of water, pumping of water is linked to high energy consumption, resulting in high recurrent costs, particularly for fuel supply and maintenance of equipment.

In addition, water supply projects are often implemented in hard to reach locations where insecurity, poor roads, and harsh weather constrain access, making the supply of fuel and other consumables for repair and maintenance expensive and cumbersome. Other inconveniences linked to the operation and maintenance of fuel-powered generators used to pump water include the risk of misuse of fuel, the logistics of ensuring regular fuel supply, and recurrent breakdowns, leading to water shortages.

The sun, on the other hand, is a potential source of energy and is environmentally friendly. Solar PV water pumping – or the direct conversion of solar

http://dx.doi.org/10.3362/9781780447810.001

energy into electricity to power a water pump – is considered in a number of contexts and regions the most appropriate and cost-effective solution to enable sustainable and reliable water supply (see Figure 1.1). Its operation and maintenance is much simpler and cost-effective, meaning that water shortages due to equipment breakdown, such as those encountered with generators, can be reduced or avoided (see Table 1.1). This makes solar PV pumping a particularly relevant technology for off-grid and weak-grid locations, and contexts where the procurement and logistics for fuel provision are too costly or erratic.

> **Solar vs wind power**
>
> Wind pumping solutions rely on wind patterns which can be highly unpredictable, require specialized parts, and must be regularly maintained, making solar pumping technology a more reliable choice for a wider number of contexts.

In particular, solar PV pumping is a relevant and cost-effective solution in vast areas of Africa, America, Oceania, and Asia, where fuel can be expensive to transport, electrical grids absent or unreliable, and access to water points limited, but where solar irradiation is fairly constant and high.

While the costs for solarizing water points are normally higher compared to diesel pumping systems, it has been established that adoption of solar PV pumping systems translates to higher savings in short to medium periods of time (GLOSWI, 2018).

Why is it then that solar pumping is not mainstreamed, especially in the relief sector? The severe shortage of solar energy expertise among NGOs in water, sanitation, and hygiene (WASH), United Nations agencies, and government water officers is probably the single most important barrier towards an effective and wider use of solar pumping technology in water supply projects, in both emergency and development contexts.

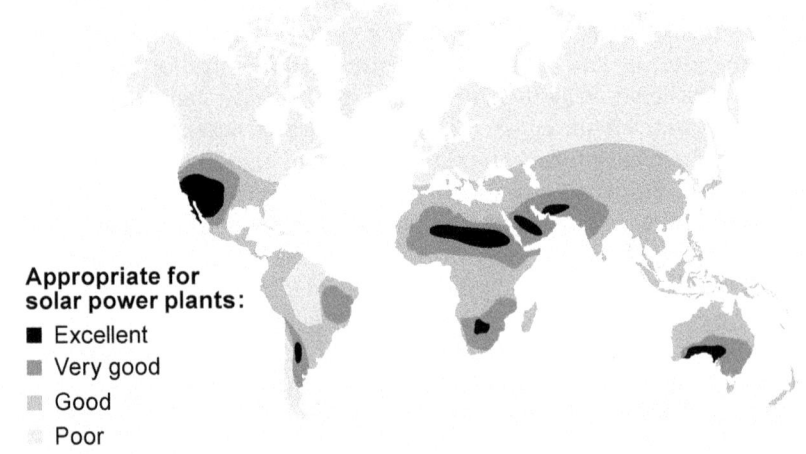

Figure 1.1 Global map of appropriate locations for solar applications

Table 1.1 Advantages and disadvantages of solar PV water pumping schemes

Advantages	Disadvantages
Low operation costs since fuel is not needed and the system is run on sunlight	Capital costs typically higher than equivalent diesel solutions. However, system prices are increasingly dropping
No dependency on erratic or expensive fuel chain supply (also avoiding the risk of fuel theft)	Most applications need water storage capacity typically larger than for equivalent diesel or grid systems
Low regular maintenance requirements since solar panels and inverters have no moving parts	Risk of theft of panels, which are seen as a valuable commodity in some locations
No pollution or noise produced	System depends on solar radiation levels
Extended lifetime (good-quality solar panels are warrantied for 25 years, inverters typically for 6–8 years)	Spare parts and knowledgeable technicians are typically available only in capital cities and are lacking at field level
A modular solution that can be expanded easily by adding modules and other accessories	Technical expertise of most humanitarian and developemnt water organizations remains low

Other constraints are that:

- Water stakeholders have not institutionalized their scattered solar expertise.
- Clear policy detailing circumstances favourable to solar pumping is lacking.
- Organizations have few incentives to adopt better energy solutions in the field.
- Practitioners are not able to properly show the benefits to donors and management.
- Some donor decisions are dominated by initial investments.

Despite these challenges, the interest in and number of solar pumping projects in humanitarian operations have increased steadily, including their inclusion in strategic documents of United Nations agencies and non-government organizations.

1.2 Factors influencing the renewed interest in solar PV water pumping

Solar pumping is not a new technology, with projects traced as far back as the late 1970s. While the possibilities of powering pumps with solar energy attracted a lot of interest initially, this faded away quickly, and it is only in recent years that a wide range of stakeholders have increasingly turned their attention to solar pumping solutions. Several factors are behind this renewed interest.

1.2.1 Environment

In the global context of climate change, the need to reduce greenhouse emissions has become of paramount importance. Environmentally friendly solutions, such as adoption of solar energy, are gaining more ground, including in emergency and development programmes.

1.2.2 Cost reduction

The use of renewable energy sources is increasing dramatically worldwide, with power installed doubling for five consecutive years to 2018 (IRENA, 2019).

Solar PV is one of the star performers, with an exponential increase over the last 10 years. In 2017 alone, over 40,000 new solar panels were installed every hour worldwide [2]. This large-scale use has resulted in economies of scale with solar PV panels (over 200-fold cheaper today when compared to the late 1970s and a cost reduction of about 80 per cent in the last ten years), with predicted further cost reductions of about 40 per cent over the next ten years.

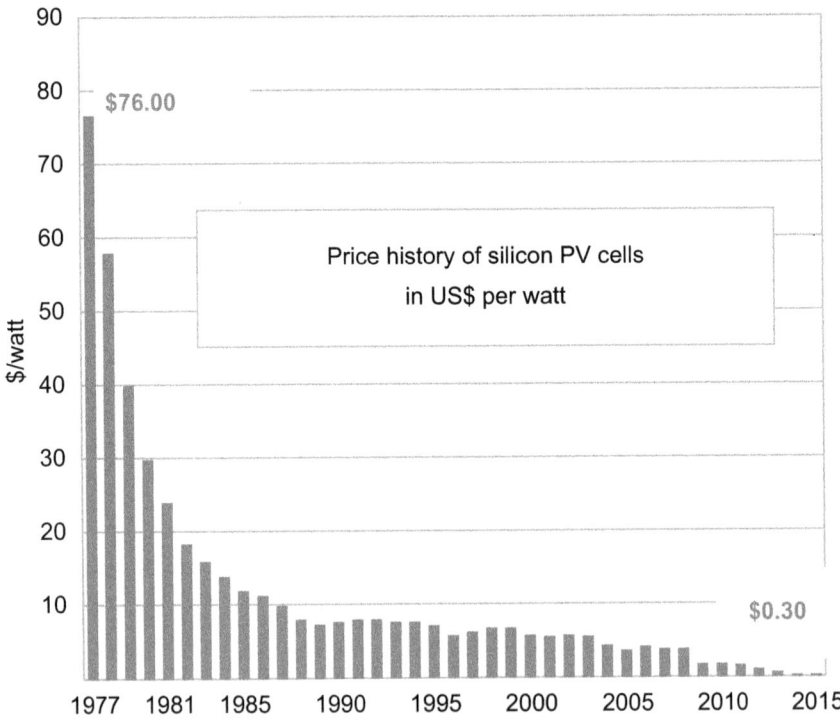

Figure 1.2 Evolution of price of PV cells, 1977–2015
Source: IEA PVPS and Bloomberg

1.2.3 High solar radiation

Areas of the world with high and regular solar radiation (radiant energy emitted by the sun), including most of the countries where emergency and development programmes take place, are also often regions that lack reliable electrical grids or where fuel is, for logistical or other reasons, expensive and/or of erratic supply (see Figure 1.1).

1.2.4 National solar private sector

Solar PV markets are rapidly evolving in the developing world with a trend that started in East Africa now spreading in West Africa and South Asia (IRENA, 2019).

This has helped create the conditions for a knowledgeable solar focused private sector to boom in many of these countries. The availability of the know-how and good-quality solar products at national level that other organizations can tap into is facilitating the expansion of the use of solar PV solutions, including solar pumping (GLOSWI, 2018a).

1.2.5 Technological advances

Technological advances in solar pumping technology have resulted in robust, versatile, and low-maintenance equipment. The extended range of powers that solar equipment can now handle means a wide range of electrical submersible and surface water pumps can be powered with solar energy.

1.2.6 National energy policies

The development of favourable policies committed to renewable energy goals from an increasing number of governments (REN21, 2019) is supporting the uptake of solar pumping solutions.

1.3 Guidance note on the use of solar pumping

1.3.1 Rationale for the use of solar PV pumping solutions

The factors mentioned in section 1.2 make conditions ripe for solar pumping to be considered as a default option for water provision in places with medium to high levels of solar irradiation (4–8 kWh/m^2), especially in off-grid locations, long-term camp contexts, or where water supply is fuel dependent but the fuel supply is too costly or erratic.

The presence of some trained private-sector contractors with good-quality solar pumping equipment in many countries further supports solar uptake and can be counted on by relief and development organizations to facilitate adoption of solar solutions for water supply projects.

There is a high potential for cost reductions to be realized if analysis and funding decisions are based on costs over the life cycle of schemes costs over time), rather than only considering capital costs of installation.

In addition, environmental considerations make solar pumping technology a climate-smart choice, especially when considered against any diesel-based option.

1.3.2 Camp contexts: mainstreaming of solar pumping

In camp contexts with the prospect of being in place for more than two to three years, solar pumping should be considered as a default solution. It should be considered from as early a stage as possible, whenever solar pumping solutions are able to meet a significant amount of the water demand (GLOSWI, 2018b).

Stand-alone solar systems should be favoured over hybrids (solar + diesel generator or other back-up power source) because of their higher cost-saving opportunity and simplicity of operation and maintenance.

> Solar pumping solutions should be the default option in refugee camps with high solar radiation and their adoption should be considered as early as possible.

However, care should be taken whenever a rapid change in context could translate into longer pumping hours going beyond the solar day, for example, when population figures are not well known or are prone to sudden increases at short notice (e.g. large refugee inflows), or when the behaviour of the aquifer is largely unknown (e.g. unknown safe pumping yields or possibility of large variations in drawdown over the seasons). In these cases, a back-up power source should be considered so that pumping beyond the solar day is possible if needed.

In older camps, solarization of water schemes should be prioritized to ensure water provision, looking first at camps with high recurrent costs or with high water shortages due to irregular or nonexistent electricity or fuel supply to power water pumps.

1.3.3 Host community context: social aspects before technology choice

Solar pumping is, from the technical point of view, equally appropriate for water supply projects at host community level (villages and towns) as for refugee or internally displaced people camps. It should be considered as a default option, in order to increase sustainability and resilience of communities. In contrast to camps where there is normally a permanent presence of relief organizations, aspects to do with ownership, operation, and maintenance add an extra layer of complexity in the host community context.

At host community level, adjustments related to the collection and use of water fees will need to be introduced and discussed, as solar solutions may not require a constant inflow of funds to operate and it may take years before equipment breaks down. Therefore, the narrative of solar-powered water systems for communities should shift from 'tapping into a cost-free source of energy to pump water' to 'accumulating funds for system replacement'.

A well-thought social approach, involving contribution from users and external technical support for maintenance and repair should come before technology choice. In this sense, prioritizing communities with strong social cohesion and coordinating approaches with government water offices and/or knowledgeable private-sector companies is a prerequisite.

1.3.4 When solar pumping should be discouraged

Solar pumping should not be seen as a blanket solution to every water supply project and its use is discouraged in some cases, namely:

- where theft and/or vandalism of solar pumping schemes is widespread, as reported from past interventions;
- when the expertise of the implementing organization is low and private-sector support cannot be counted upon;
- where solar technology does not bring any significant technical, economic, or enviromental advantage over existing solutions in terms of amount of water supplied, greenhouse gas emission reduction, cost savings over time, or simplicity of operation and maintenance of equipment.

1.3.5 Issues related to operation and maintenance, and training and evaluation

Solar pumping schemes will suffer fewer breakdowns and have much less intensive maintenance than generator or hand pump schemes. However, solar pumping schemes can and will experience technical problems at some point in time that cannot be solved at community level (or for which the organization in charge of the water scheme will probably need external support), regardless of the training provided in the past.

> Geographical clustering and maintenance service agreements are a good way to ensure timely servicing and repairs in places where parts and technicians are available only in capital cities.

It is important that service agreements are established with a good-quality private contractor, water utility, water service provider, or relevant government technical office, before any installation and they should be renewed as needed.

Since the single most important barrier towards a wider adoption of solar pumping solutions is the weak technical expertise of most WASH organizations, support from the private sector, government, and/or the donor community should be provided or encouraged for capacity-building activities in areas with high potential for adoption of solar pumping solutions (e.g. areas with high solar radiation and high dependency on fuel-based solutions for the supply of water).

In addition, ways of collaborating (with, for example, academic institutions, knowledgeable private companies, or water utilities) should be encouraged in order to make training activities as sustainable or over as long a duration as possible.

Finally, since adoption of solar pumping solutions is often based on the long life expectancy of solar products, it is important to support evaluations of older (more than five years) solar systems in order to build up stronger evidence on the adequacy of solar pumping technology for the given context, as well as to inform future water strategies in the country of work.

Facts and figures

- The working life of solar modules is warrantied to 25 years.
- The cost of solar modules has decreased by 80 per cent in the last 10 years.
- Cost recovery of solar pumping investments vs diesel technology is on average 0 to 4 years.
- Cost reduction over life of the system is 40–90 per cent when compared to diesel generators.
- Well-designed and maintained solar systems can function for more than 10 years without any major failure.
- Solar pumping heads of +450 m; pumping rates of +240 m^3/h.
- Average cost of service agreements with contractors to ensure functionality is about US$1,500 a year.

Source: GLOSWI, 2018

CHAPTER 2
Definitions and principles of solar energy production

The high and constant level of solar irradiance in most areas where relief projects take place make solar photovoltaic pumping an ideal choice for water supply projects. Sun and water work well together as the sunnier it is, the more electricity is produced and the more water is pumped. A number of basic solar and electricity concepts need to be understood by water engineers in order to feel confident in the use of solar PV solutions for their water supply projects, including the different electrical parameters of a solar module, the parameters to measure the solar resource in a certain area, and basic DC electricity concepts and formulas.

Keywords: solar resource, irradiation, insolation, I-V curve, maximum power point tracking (MPPT), solar maps, solar DC, peak sun hours (PSH), solar photovoltaic, standard test conditions (STC), solar module, PV array

2.1 The solar resource

Globally, energy consumption is expected to double by 2050 (IRENA, 2019). Oil supplies are increasingly dwindling and solar provides a clean and viable renewable energy alternative for meeting the growing energy demand while mitigating against environmental damage caused by fossil fuel use. Solar energy is renewable because it will be available as long as the sun continues to shine. The sun is estimated to provide the same level of solar radiation for another 4 to 5 billion years.

The enormous amount of energy provided by the sun can be tapped directly using photovoltaic technology. More energy from sunlight strikes the earth in 1 hour (4.3×10^{20} J) than all the energy consumed on the planet in a year (4.1×10^{20} J). The energy from the sun is free, sustainable (cannot be depleted), is found everywhere, and is non-polluting. Yet solar energy comes with its own shortfalls as it is not constant during the day and through the year. There are, however, ways to overcome these shortfalls, as will be seen through the design process of a solar water pumping scheme.

The abundant solar energy available in most regions of the developing world can be exploited for water pumping for both drinking water and irrigation, particularly in humanitarian and development contexts which are often faced with severe water and energy supply challenges.

2.2 Sun and water: the perfect relationship

Water for drinking, irrigation, livestock watering, and other purposes is obtained from a variety of surface sources (e.g. rivers, lakes, dams, water pans, berkads, springs, rock catchments) and underground sources (e.g. shallow wells, deep boreholes).

Energy is required to pump the water from the source to the point of use. Visits by the Global Solar and Water Initiative (2016–2020) to more than 100 camps and communities in 12 countries, which are reported in GLOSWI country reports, have revealed that the majority of water supplies are from underground sources, particularly deep boreholes. Pumping from deep boreholes requires relatively high amounts of energy, which is often lacking in many project locations, particularly, where grid electricity is non-existent or unreliable and transport of fuel expensive and difficult.

The abundance of solar energy, the great need for water, and the lack of other energy sources in these locations introduces a perfect relationship between the sun and water. In other words, more sun means more water is required. Surface waters decrease in the dry season, whilst arid lands have greater water needs and agricultural water needs increase during the dry season. More sun also means more solar power is available, and more power means more water can be pumped. This is a natural correlation that bodes well for PV water pumping.

With well-established, sophisticated technologies, solar energy is used to power a water pump that transfers water from the source to the point of delivery. As long as there is enough sunshine, water can be pumped

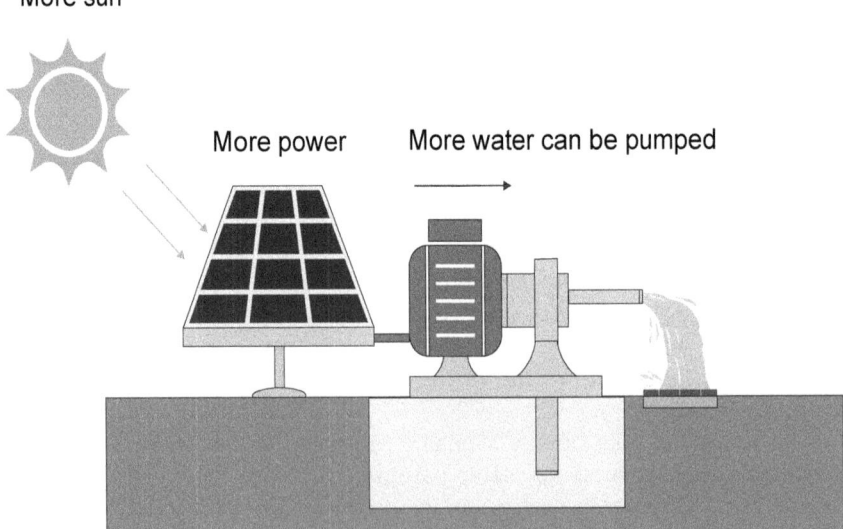

Figure 2.1 Sun and water: a perfect relationship

either to a tank or directly to the consumer's tap, keeping the arrangement simple, efficient, and free of energy storage devices, such as batteries.

2.3 Solar radiation

Solar radiation is a general term for the electromagnetic waves emitted by the sun. In space, solar radiation is practically constant; on earth, it varies with the day of the year, time of the day, the latitude, and the state of the atmosphere (e.g. presence of humidity, smoke, smog, dust), but remains predictable and constant year to year. A solar surface known as a solar collector captures the solar radiation, with the position of the surface and the local landscape also influencing the amount of radiation that can be collected.

Solar radiation can be converted into electricity using photovoltaic technologies, such as PV modules. The amount of electricity generated from a module depends on the sunlight intensity. The more intense the sunlight, the more electricity is produced.

As sunlight passes through the atmosphere, some of it is absorbed, scattered, and reflected by air molecules, water vapour, clouds, dust, and pollutants. The total radiation received on a horizontal surface on the earth is called the Global Horizontal Radiation and is made up of direct, diffuse, and reflected radiation. Diffuse solar radiation is the portion scattering downwards from the atmosphere that arrives at the earth's surface. The reflected portion is the energy that bounces off from the surface of the earth. The solar radiation that reaches the earth's surface without being diffused in the atmosphere is called direct solar radiation (Figure 2.2).

Measurements of solar energy are typically expressed as total solar radiation on a horizontal or tilted surface and calculated from the relationship;

Global Horizontal Radiation = direct + diffuse + reflected

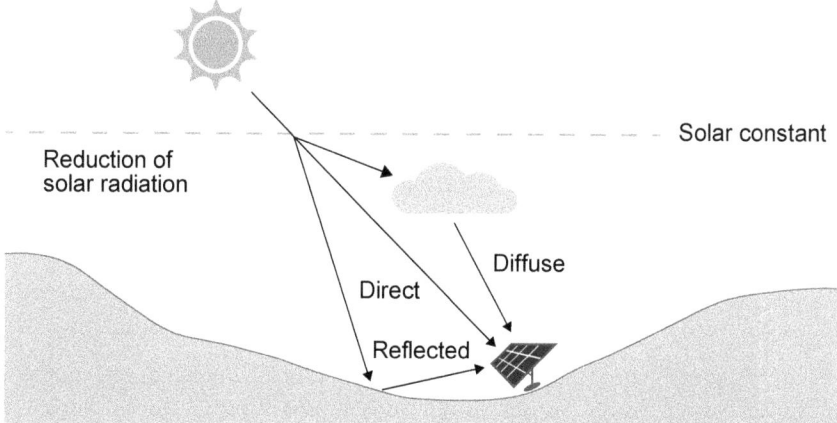

Figure 2.2 Global Horizontal Irradiation

In designing and sizing solar energy systems, the quantification of the amount of solar energy incoming to solar collectors can be represented as irradiance and insolation (discussed in section 2.5 and section 2.6).

2.4 Solar photovoltaic

The term solar photovoltaic (PV) is used to refer to generation of electricity from the sun's energy. A solar PV cell transforms the energy of the sun into electricity. The PV effect happens when photons of light hit a collection of solar cells, exciting electrons into a higher state of energy, making them act as charge carriers of an electric current. Explained differently, solar radiation strikes solar cells, the atoms in the solar cells absorb some of the photons of sunlight, get excited and release electrons. The electrons flow through a conductor to produce electrical current. The current is proportional to the intensity of the radiation striking the collector (Figure 2.3).

The PV effect takes place in a semiconductor material, the most widely used being crystalline silicon. The active semiconductor material (silicon) has atoms that are held together by valence electrons. Since silicon has very few electrons at room temperature, it is 'doped' to create a positive-negative junction (p-n junction). This is done by introducing particular impurities, such as boron (p-type) which has an excess of holes (absence of an electron is a hole that leaves a net positive charge), or phosphorus (n-type) which has an excess of electrons (Figure 2.4). This makes the semiconductor material more conductive (carries current), provides an electric field (creates voltage), and allows current to flow only in one direction. The valence electrons

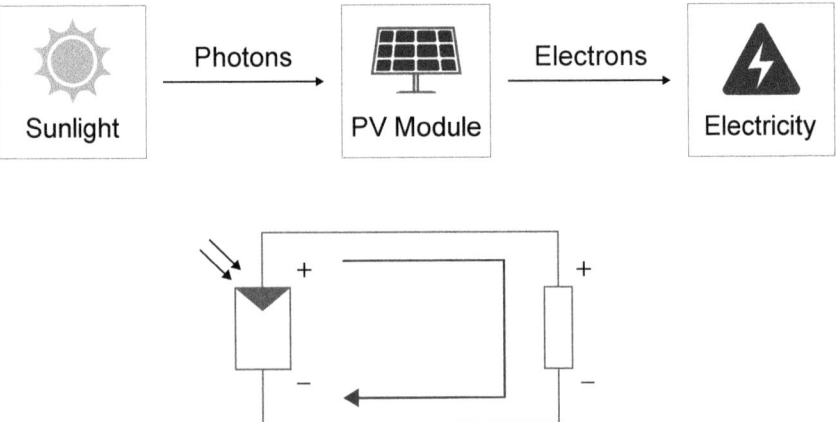

Figure 2.3 Conversion of sunlight to electricity

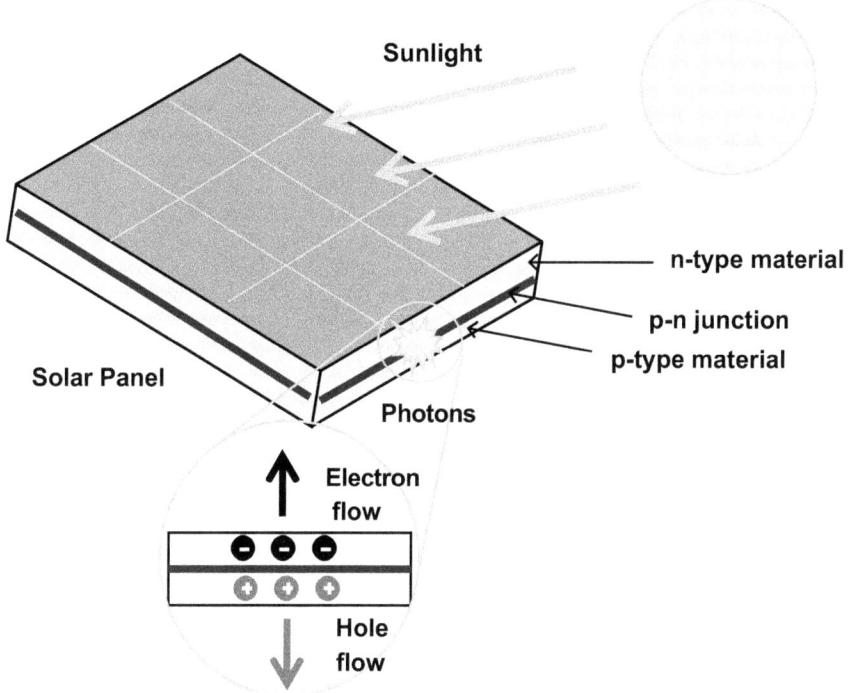

Figure 2.4 The 'PV effect' process

Box 2.1 Solar PV components

Solar cell – The basic solar building block, an individual photovoltaic cell. They convert light to electricity.

PV module – Also known as a solar panel, PV panel, or solar module, this is a collection of interconnected solar cells usually encased in glass with an aluminium frame.

Solar array – Several modules electrically connected in series or parallel.

Solar generator – Several arrays electrically connected in series and parallel to increase the total available power to the required voltage and current.

String – A set of solar modules electrically connected in series.

---

are released by energizing them using light and they become free to move, creating electricity.

A single cell will only produce a small current or voltage, unsuitable for most applications. Therefore, cells are added together in various series or parallel configurations, and encapsulated in aluminium or glass to produce solar modules with varying current, voltage, and power outputs.

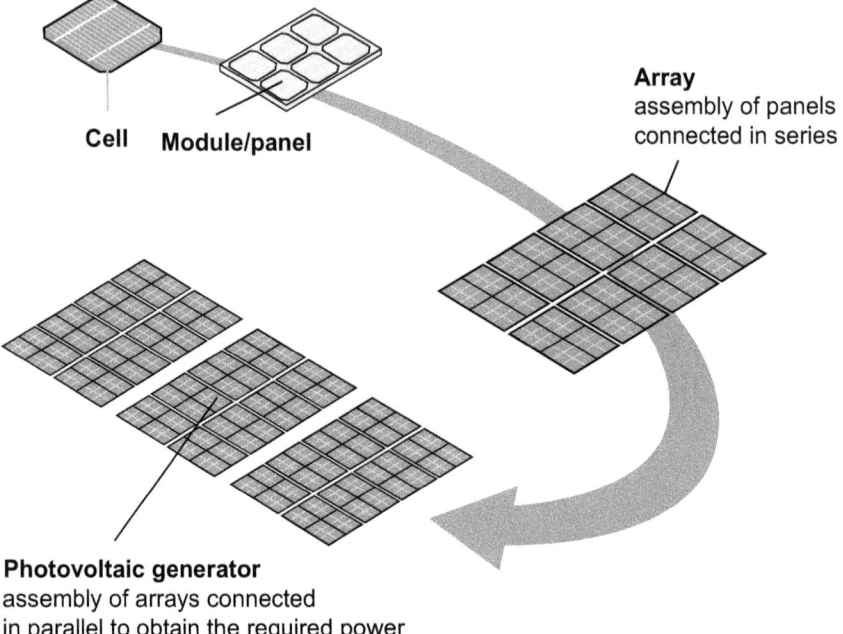

Figure 2.5 Illustration of PV cell, module, array, and generator

The solar PV generator is connected through a control unit (controller) to power a pumping system (the load). The basic configuration of the system consists of a PV array/generator, a controller, and an electric pump (see Figure 2.6).

2.5. Solar irradiance

Solar irradiance is used to define the instantaneous power of solar radiation received on a surface per unit area. It is expressed in W/m^2 (or kW/m^2). The typical amount of irradiance received on a tilted surface at the equator on a clear day at noon (when the sun is overhead) is equal to 1,000 W/m^2. This means 1,000 W is received on a 1 m^2 surface at any given time. On clear days, a surface receiving solar energy will capture mainly direct radiation, while on cloudy days the surface will receive mostly diffuse radiation because direct radiation has been obstructed by the clouds. Irradiance values greater than 1,000 W/m^2 can also be achieved due to a combination of direct and diffuse radiation.

Non-tropic countries (e.g. countries in Europe) experience lower solar radiation than equatorial countries because the incident angle of the sun's rays is not perpendicular to the earth, and much more of the resource is absorbed in the atmosphere.

DEFINITIONS AND PRINCIPLES OF SOLAR ENERGY PRODUCTION 15

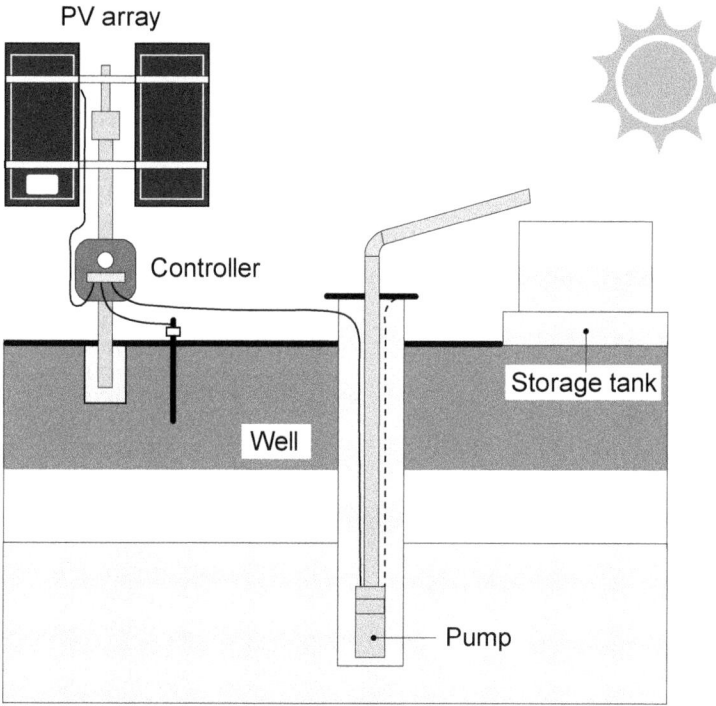

Figure 2.6 A typical solar-powered water system

The solar irradiance will vary throughout the day with minimum values at dawn and dusk and maximum values at midday. For example, on a clear day in Valencia, Spain, the irradiance value at 9:00 a.m. will be less than the irradiance value at noon. This is explained by the earth's rotation about its axis, which causes the distance travelled by sunlight through the earth's atmosphere to be at a minimum at solar noon. At this hour, the sun's rays are striking a surface perpendicularly and through the least atmosphere.

Figure 2.7 shows the solar radiation arriving on a photovoltaic installation in Valencia, oriented to the south and with a tilt of 30 degrees with respect to the horizontal during a week in the middle of July (summer). The maximum level of solar radiation is 900 W/m² at noon.

Figure 2.8 shows the changes in the amount of solar energy received on a surface during a clear day (left part of the plot). In the morning and late afternoon, less power (irradiance) is received because the surface is not at an optimum angle to the sun. At noon, the amount of power received is the highest (around 900 W/m²). However, the actual amount of power received instantaneously varies with passing clouds and atmospheric clarity due to dust in the atmosphere (the right part of the plot shows the effect of some passing clouds at noon).

16 SOLAR PUMPING FOR WATER SUPPLY

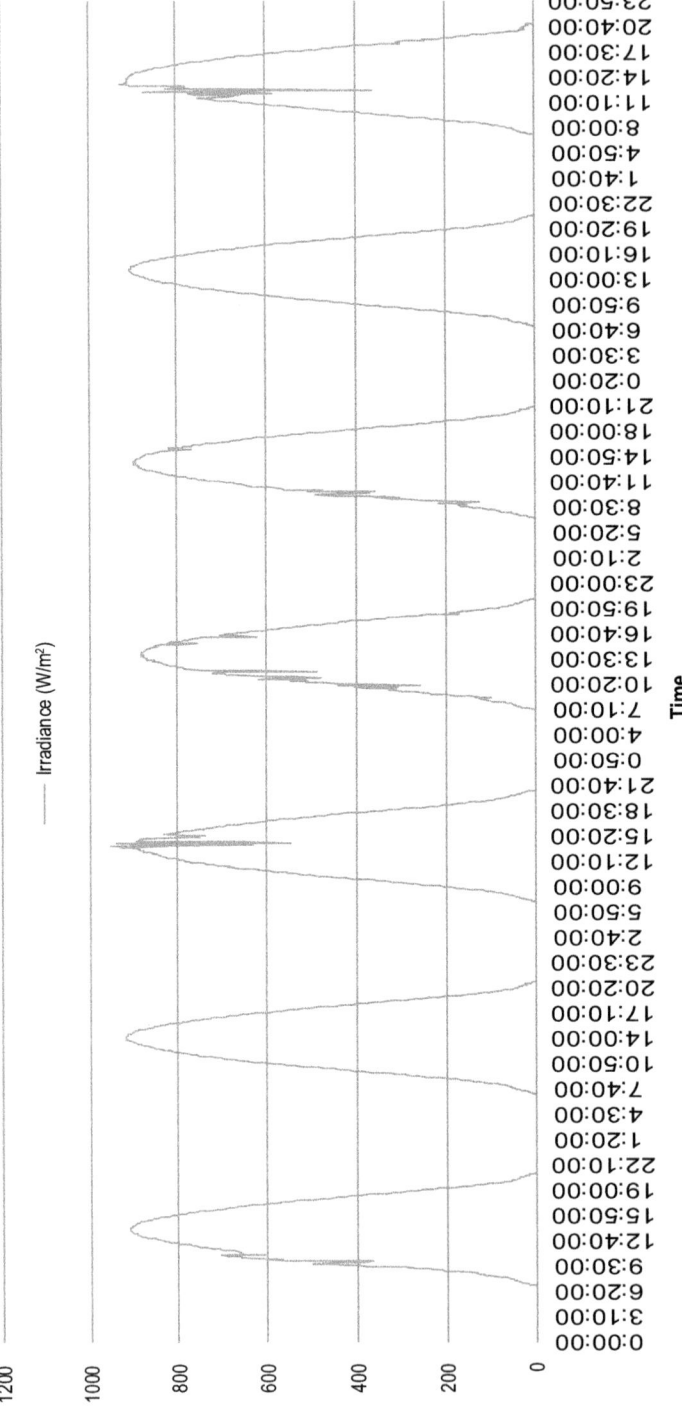

Figure 2.7 Weekly irradiation in Valencia in the month of July
Source: Polytechnic University of Valence, Spain

DEFINITIONS AND PRINCIPLES OF SOLAR ENERGY PRODUCTION 17

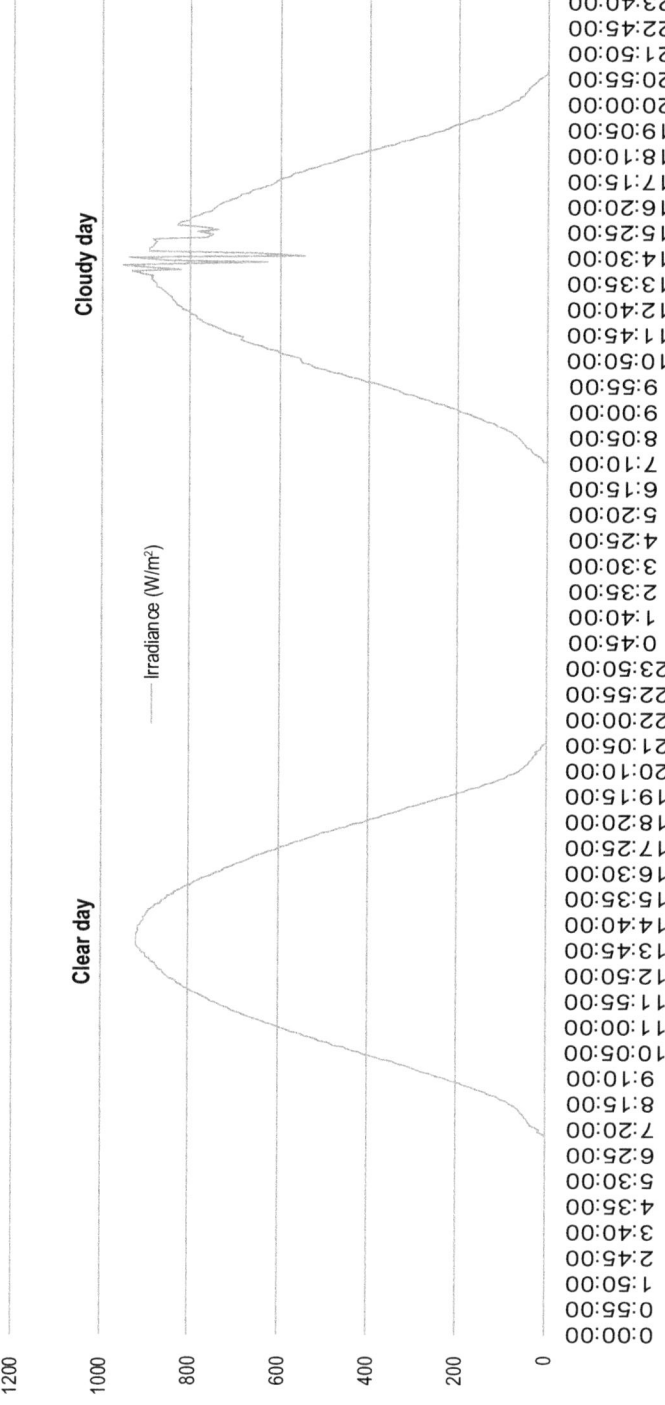

Figure 2.8 Time-based solar irradiance in Valencia during summer for two consecutive days
Source: Polytechnic University of Valence, Spain

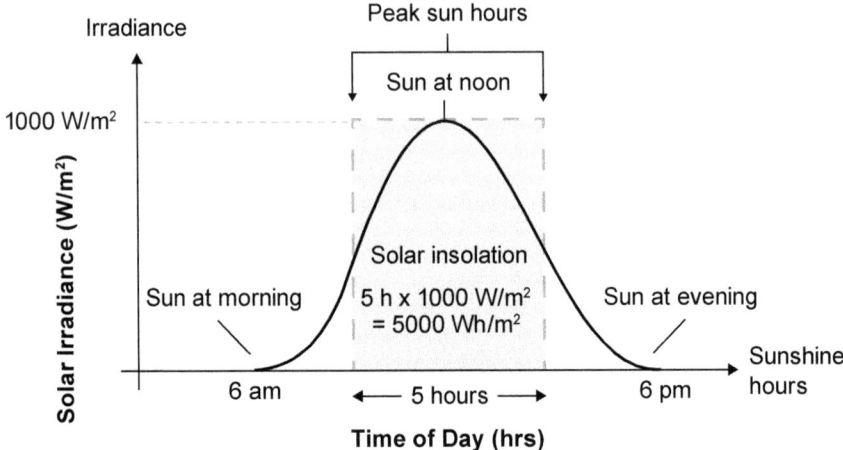

Figure 2.9 Graph of daily insolation

2.6 Solar insolation

Solar insolation is a measure of the cumulative irradiance received on a specific area over a period of time, measured in Wh/m^2 or kWh/m^2. It is typically the area under the irradiance curve over hourly, daily, monthly, and yearly data. For practical design purposes, the daily insolation is considered, that is, the accumulation of irradiance over a day. Simply put, it is the energy received on a 1 m^2 surface over a period of one hour, hence the unit of kilowatt hours per square metre (kWh/m^2).

Insolation is also expressed in terms of peak sun hours or the number of hours per day during which solar irradiance averages 1,000 W/m^2 (see Figure 2.9). For example, if the daily insolation is 5 kWh/m^2, then dividing by 1 kW/m^2 (1,000 W/m^2) gives five peak sun hours.

As can be seen in Figure 2.10, the irradiance changes throughout the day. The energy of the first day is equivalent to seven hours of irradiance at 1,000 W/m^2. Seven peak sun hours (PSH) means that the energy received during total daylight hours equals the energy that would have been received if the sun had shone for seven hours at a rate of 1,000 W/m^2. The second day has 6.6 PSH due to the smaller irradiance during the passing of the clouds. Peak sun hours are useful because they simplify design calculations. The energy produced by a PV array is directly proportional to the amount of insolation received.

2.7 Standard test conditions

The disadvantage of solar-powered systems is that energy supply is not continuous and constant during the day and varies from day to day throughout the year. The amount of energy received will vary depending

DEFINITIONS AND PRINCIPLES OF SOLAR ENERGY PRODUCTION 19

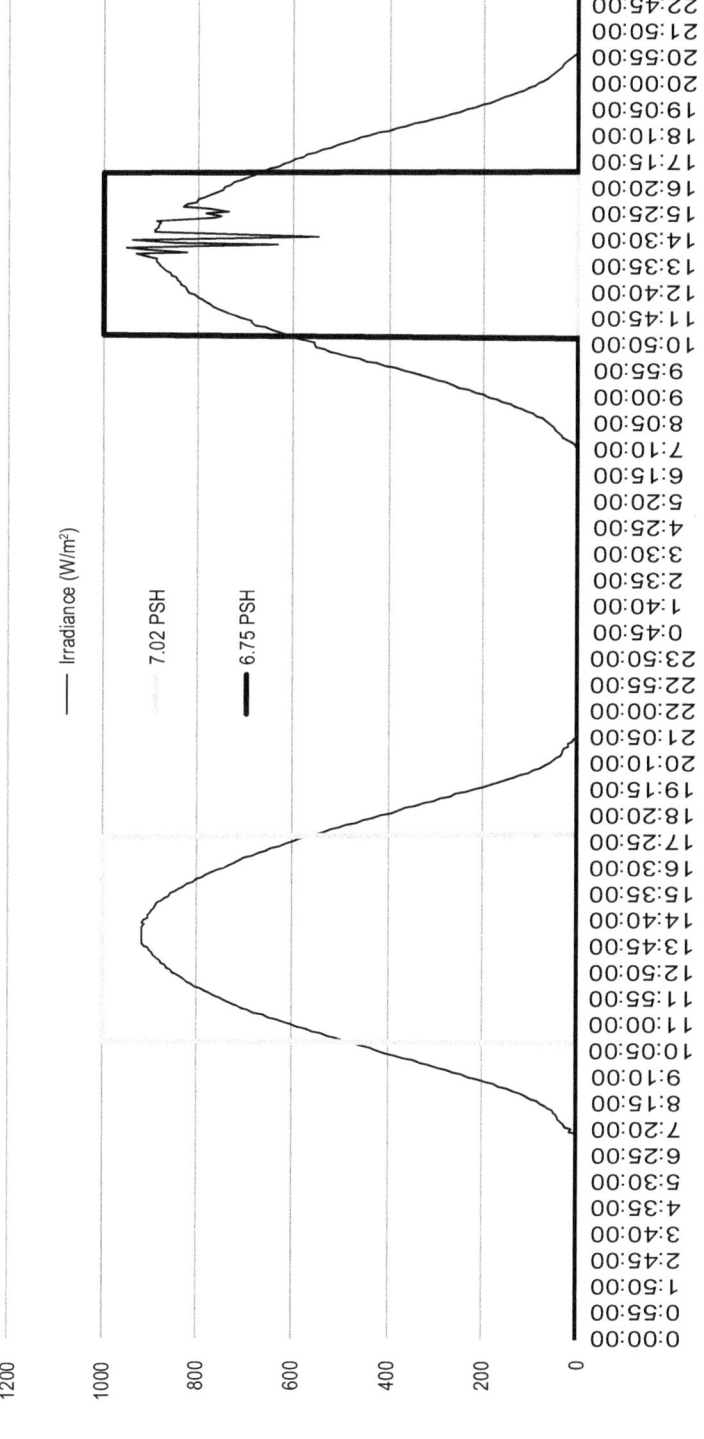

Figure 2.10 Solar irradiance during two days of July and corresponding insolation
Source: Polytechnic University of Valence, Spain

on the location, the season, the time of day, and the weather (especially cloud cover). This will in turn affect the amount of power produced by a solar module and hence the water output in a solar-powered water system. Therefore, a PV module will give different power outputs at different locations, different seasons, different times of the day, and different weather conditions.

For uniformity, PV modules are tested and rated in standard test conditions. STC make it possible to conduct uniform comparisons of photovoltaic modules from different manufacturers and to accurately compare and rate them against each other.

Solar modules are tested and rated at STC of:

- solar radiation of 1,000 W/m² – typical at noon on a clear day
- module temperature of 25 degrees Celsius – the temperature of the cell itself and not ambient temperature
- air mass equal to 1.5 (AM1.5) – the thickness of atmosphere the sun passes through. AM0 is the value of solar radiation outside the earth's atmosphere (1,350 W/m²). AM1 is the value on the earth's surface when the sun is overhead and the radiation travels through the thickness of the atmosphere at a right angle (Figure 2.11).

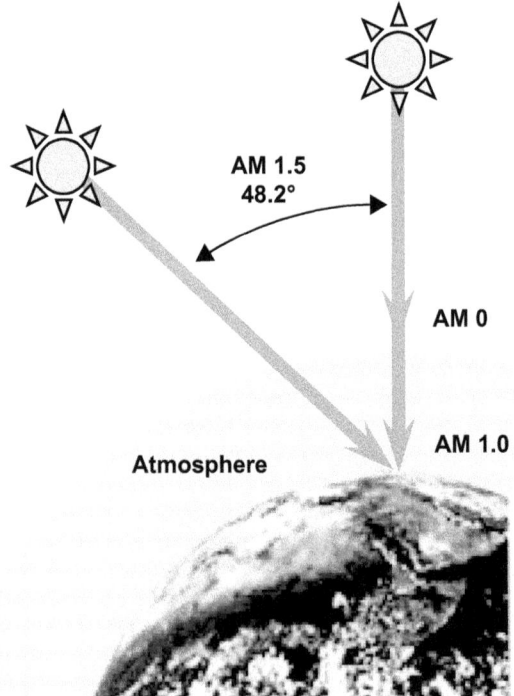

Figure 2.11 Air mass for different sun positions

A peak watt (Wp) is the amount of power output a PV module produces at STC. When a module is working outside of STC conditions (which is often the case in real-life conditions), output will vary according to the prevailing conditions of irradiation and temperature (air mass has little effect in practice). A module with a rating of 300 W_p will give only 300 W at 1,000 W/m^2 solar irradiation and 25°C cell temperature. When the irradiation drops below 1,000 W/m^2 and/or the cell temperature rises above 25°C there is a corresponding decrease in the power output from the module. It is possible for this module to give more than 300 W when the irradiation is more than 1,000 W/m^2 and/or the cell temperature below 25°C within the tolerance value stated by the manufacturer. This is discussed in more detail in Chapter 4.

In theory, a water pump of 3 kW connected to panels rated at 3 kW under STC would run as if connected to a generator or grid power.

2.8 Solar resource maps for peak sun hours

Peak sun hours are used to size and design a suitable PV generator to meet the energy demand of the pump to provide the required water output. PSH is used to define the amount of energy available per day in a given location.

The PSH is a useful value for comparing the energy differences at different locations. In pumping, the PSH is used to give an indication of how many hours in a day there will be maximum power for the pump to achieve peak pumping.

Databases are available that use systematically processed historical weather information of many years to provide long-term average weather data. Three of these databases are:

- Solargis solar resource maps <https://solargis.com/maps-and-gis-data/download/>
- POWER Data Access Viewer <https://power.larc.nasa.gov/data-access-viewer/>
- European Commission's Photovoltaic Geographical Information System <https://re.jrc.ec.europa.eu/pvg_download/map_index.html>

These three databases are reputable sources of high-quality, reliable solar resource data that is validated globally, and one can find data for most locations in the world. The maps or data can be downloaded for free by region or by country (terms of use can be checked on these websites). The global solar irradiation map shown in Figure 1.1 gives information about the annual and daily PSH (kWh/m2/year or day) in the different regions of the world.

Looking at the map in Figure 1.1, Africa has approximate daily PSH values of between 4.6 and 7.4 hours, while Europe has daily PSH values from as low

Example of PSH and pumping

After a pumping test, a new borehole in Madaba, Jordan, has a safe yield of 20 m³/hr. The borehole will supply a village with a population of 2,000 people at 40 litres per day for each person. Ignoring temperature effect and other factors, can this demand be met using solar pumping?

2,000 people x 40 litres/day = 80,000 litres/day = 80 m³/day

From the solar resource map for Jordan (see Figure 2.12), Madaba has a yearly insolation of about 2,100 kWh/m². This translates to a daily insolation of 5.8 kWh/m² (i.e. 2,100/365 = 5.8). This means that the peak sun hours for Madaba is 5.8 PSH (i.e. Madaba receives 1,000 W/m² over a translated period of 5.8 hours daily).

Maximum possible borehole supply = safe yield x PSH = 20 m³/hr x 5.8 = 116 m³/day

The amount of water that can be pumped during the peak sun hours is more than the demand. It is possible to supply the water needs of this village with a solar scheme.

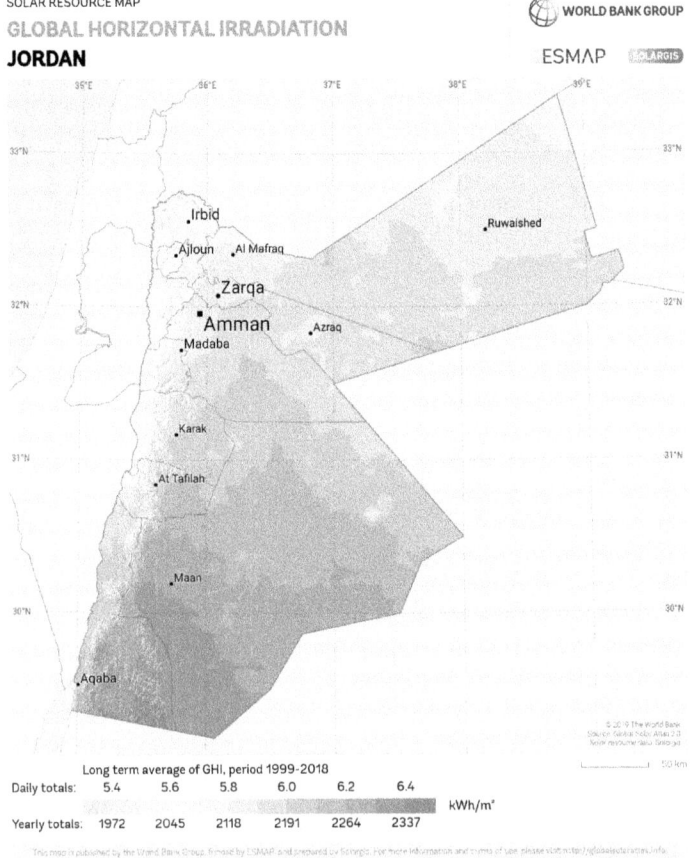

Figure 2.12 PSH map for Jordan
Source: solargis.com

as 2.2 up to 4.6 hours. In theory this means that at STC, a pump connected to a PV power source can produce peak flow for 7.4 hours in some parts of Africa but will only produce peak flow for a maximum 4.6 hours in some parts of Europe (assuming the same size of PV power and same installation in both scenarios).

In practice, it will be seen in the following chapters that by oversizing the solar PV generator in a water pumping scheme, longer hours of pumping can be achieved.

To get the global irradiation map of any region or country, visit https://solargis.com/maps-and-gis-data/download, from the drop-down menu select region (e.g. Africa, Asia), then select country (e.g. Kenya, Jordan), scroll down to Global Horizontal Irradiation and click on download.

> **Application note**
>
> When the PSH is less than 3.0 hours, it would be important to critically examine the cost of installing solar PV pumping vis-a-vis the benefit to be realized. It may be necessary to discard the solar option for such low PSH values unless it is the only option available.

2.9 Basic DC electric concepts

Current, voltage, and resistance are three important concepts of electricity it is important to become familiar with to understand power generation using solar.

A small light bulb connected to the positive and negative side of a solar module creates a closed circuit where electrons can flow between the terminals (through the wires) to make the lamp light up. This flow of electrons is called *electrical current*.

In this circuit, there will be an opposition to the flow of electrons and it is called *resistance*.

Current will only flow if there is an electrical potential difference between the positive and negative terminals. The potential difference between the terminals is called *electrical voltage* and is caused by the resistance to flow of electrons. It is the force exerted on the electrons to produce current.

Ohm's law relates the three as follows:

| Voltage (V) | = | Current (I) | × | Resistance (R) |

The *electrical power* is the product of current and voltage and electrical power is what is generated at any given instant. The relationship between power, current, and voltage is expressed using the *power law*.

| Power (P) | = | Current (I) | × | Voltage (V) |

The *electrical energy* is the power generated during a period of time.

This concept can also be understood in this manner: the flow of water in a pipe is the equivalent of the current in an electrical circuit. For the water to flow through the pipe there must be a pressure difference (head) between the two points, and this pressure difference is equivalent to voltage. If the pipe is clogged at some point or due to friction, the flow of water is restricted and the flow through the pipe is reduced. The friction (or clog) is equivalent to resistance. This is summarised in Table 2.1.

Power losses in a cable are influenced by the length, size, and type of wire conductor. Specifically, resistance is directly proportional to the length of the wire and inversely proportional to the thickness of the wire. In other words, the longer the wire, the greater the loss and the larger the wire diameter, the less the loss. The energy loss is also influenced by the wire material: a good conductor, such as copper, has a low resistance and will result in less energy loss.

Table 2.1 Analogy of electricity and water

Electricity in a cable			Water in a pipe		
Definition	Term	Unit	Definition	Term	Unit
Flow of electrons	Current (I)	Amperes (A)	Flow of water	Flow (Q)	m³/hr
Potential difference	Voltage (V)	Volts (V)	Pressure difference	Head (H)	Metres (m)
Resistance to flow	Resistance (R)	Ohms (Ω)	Resistance to flow	Friction (H_f)	Metres (m)
Instant power generated = I x V	Electric Power (P)	Watts (W)	Hydraulic load = Q x H	Hydraulic power (P)	Watts (W)
Power generated in a period = P x time	Electric Energy (E)	Watt hours (Wh)	Power generated in a given time	Hydraulic energy (E)	Watt hours (Wh)
High voltage + small cable = High current + high resistance + cable losses = heat and fire			High pressure + small pipe = High velocity + high friction losses = pipe bursts		

Another effective way to reduce electrical losses in a system is to decrease the current flow (keeping the electric power of the application). Power losses in an electrical circuit are proportional to the square of the current, as shown in this equation.

$$\text{Power loss} = \text{Current}^2 \times \text{Resistance}$$

Consequently, from the power law (P = I × V), increasing the voltage while reducing the current will result in the same power transmission, but with less power loss. Therefore, higher voltage pumps tend to be more efficient than lower voltage pumps, assuming all other properties are similar.

It is important to keep in mind that there are two forms of electricity, *alternating current* (AC) and *direct current* (DC), and the two cannot be mixed together directly. Solar PV modules always produce DC power, while fuel generators and grid sources give AC power. Direct current has fixed polarity (i.e. does not change direction) but for alternating current the polarity changes. A DC load (such as a DC pump) is powered directly using DC power through a simple on–off switch. An AC load (such as an AC pump) can also be powered using DC power from the solar modules, but the DC power must first be converted to AC to power the AC load. This conversion is achieved by use of a DC–AC inverter. More on this is discussed in Chapter 3.

2.10 Solar module I-V curve and maximum power point

The electrical characteristics of a PV module can be defined using the relationship between the current and voltage plotted on a curve (the same way that the hydraulic characteristics of a water pump are defined by plotting on a curve their hydraulic head versus the flow output). The values of current and voltage vary from zero to a maximum and are obtained by exposing the module to a constant irradiation and temperature, varying the load resistance from zero to infinity and measuring the current and voltage. The values are then plotted on an I-V curve on the horizontal and vertical axes respectively, as shown in Figure 2.13.

Open-circuit voltage (V_{oc}) is the maximum voltage measured and is achieved when there is no load connected to the module (open circuit) meaning there is a potential difference (voltage) but no flow of electrons (zero current). In solar design, the V_{oc} is used to calculate the maximum voltage input into the controller.

Short-circuit current (I_{sc}) is the maximum current and is achieved when the circuit is shorted, meaning there is a flow of electrons but no potential difference (zero voltage between the negative and positive terminal). In design,

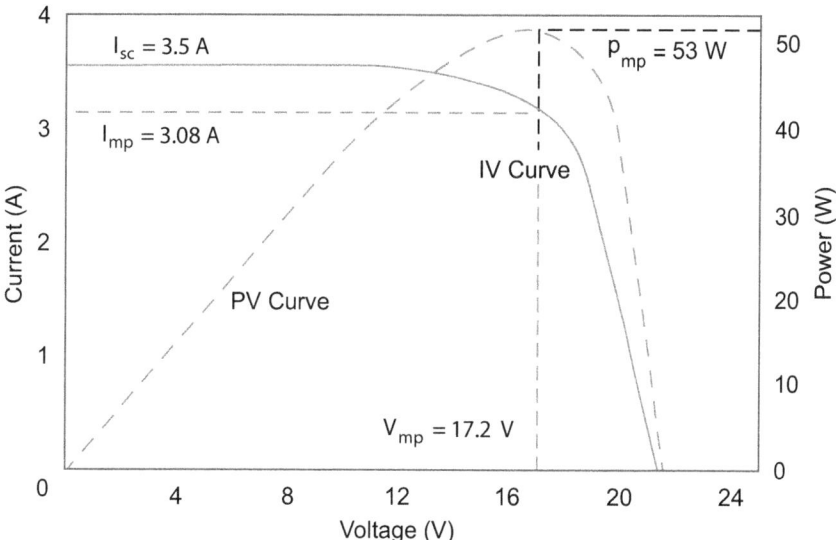

Figure 2.13 Typical I-V and power curves for a crystalline PV module operating at STC

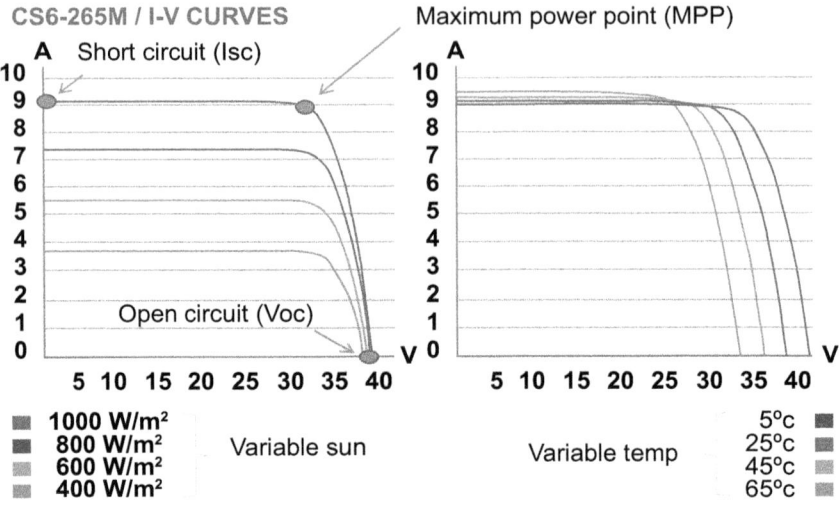

Figure 2.14 I-V curve under varying irradiance and temperature

the I_{sc} is used to size all the DC accessories, including DC switches and DC cabling, since it is the maximum current generated by the module.

Since practically the two maximum values cannot be achieved on load (i.e. no power is produced at short-circuit current with zero voltage and at open-circuit voltage with zero current), the maximum power is obtained at the point on the curve where the *product of voltage and current is maximum*. This occurs at the knee of the curve. This point is known as the maximum power point (MPP) and is the highest possible power output of a solar module. This point represents the maximum efficiency of the module in converting sunlight into electricity. The point at which power is maximum gives the *maximum peak current* (I_{mp} or I_{MPP}) and the *maximum peak voltage* (V_{mp} or V_{MPP}). The MPP is the desired point of operation of the module and operation outside this point reduces the amount of power generated.

In real conditions, irradiation and temperature are not constant and therefore the module's MPP will shift according to changes in irradiation and temperature (see Figure 2.14). Lower irradiances reduce the current output but its effect on voltage is negligible, and higher temperatures reduce the voltage but its effect on current is minimal.

Since power is the product of current and voltage (Ohm's law), then reduction in irradiance and increase in temperature results in reduced power output. By using MPP tracking (MPPT) technology in solar controllers, the resistance of the load can be varied to ensure that the module is always operating at its maximum power point at varying current and voltage conditions. Solar pump controllers will therefore include an MPPT function that will take power out of the PV array at its most efficient value by adjusting the load, thereby avoiding needless energy losses. MPPT is an important feature to look out for in solar controllers.

CHAPTER 3
Solar-powered water system configurations and components

While solar pumping is not a new concept, with projects dating back to the late 1970s, the technical revolution in terms of inverters (converters of solar DC electricity into AC) has opened the door to the solarization of a much wider range of water pumps, both surface and submersible. Depending on whether the water required can be supplied during the solar day or pumping is needed beyond, configuration of schemes will be stand-alone or hybrid (one or more energy sources). This chapter presents the main components of a solar pumping scheme.

Keywords: balance of system components, inverters, solar stand-alone, solar hybrid, solar pumps, solar controllers, variable frequency, disconnect switch

3.1 SPWS concept and revolution

The benefits of solar-powered water systems (SPWSs) cannot be overemphasized: simple, reliable, durable, modular, and low maintenance. In SPWSs, the solar energy is coupled directly to power an electric pump motor through a solar controller. The electric pump can be either a surface pump, submersible pump, DC pump, or AC pump. The controller can be either a DC control box or an inverter. Therefore a photovoltaic water pumping system is generally like any other pumping system, with the exception that the power source is solar energy.

Previously the capacity of SPWSs was limited. Fifteen years ago, the biggest solar pump on the market was probably a 4 kW DC pump with a daily hydraulic duty of between 1,500 and 2,000 m^4 (approximately 10–200 m^3/day at an inverse head of 10–200 m). The drastic reduction in solar PV prices in the last decade has triggered technological advancements of robust and reliable solar pumping equipment.

The development of variable-frequency inverters has extended the solar pump performance range tenfold since they work with standard electric motors. Literally any electric pump can be solarized and can also be powered using dual power sources (solar and AC power). These developments have led to a revolution in the use of solar for off-grid water pumping and the emergence of a vibrant private sector with good technical knowledge offering quality solar pumping products in most countries (see GLOSWI country reports from 2016 to 2020). SPWSs are technically non-restricting these days with large solar pumping systems feasible (see more on SPWS sizes in section 7.5).

http://dx.doi.org/10.3362/9781780447810.003

SOLAR PUMPING FOR WATER SUPPLY

Table 3.1 List of solar-powered water systems found in the field

Country	Location	Motor size (kW)	PV generator size (W)	Average output (m³/day)	Total dynamic head [m]	Price (USD)[1]
Kenya	Lodwar	0.9	1,400	18	70	$8,535
Nigeria	Maiduguri	1.4	1,000	17	61	$7,955
Sudan	Zamzam	1.7	1,560	26	55	$11,013
Ethiopia	Fafen/Awbare	3.0	7,500	108	47	$24,309
South Sudan	Bentiu	4.0	10,920	120	68	$23,998
Uganda	Rhino	4.0	4,000	30	90	$17,057
Ethiopia	Fafen/Babile	4.0	6,750	59	80	$20,264
Kenya	Songot	4.0	3,500	35	85	$15,893
South Sudan	Bentiu	5.5	15,600	197	58	$32,747
Uganda	Bidibidi Zone 1	7.5	15,000	95	121	$34,554
Iraq	Erbil	18.5	32,130	277	110	$50,567
Kenya	Nablon	30.0	51,000	450	100	$79,820

Note: [1] The price is for an installed SPWS up to the borehole surface, including pump, motor, solar generator, solar controller, DC accessories, cabling, and drop pipes.
Source: GLOSWI Country reports, 2016–2020

Table 3.1 shows a representative list of some of the systems encountered in the field together with their prices.

One of the distinguishing factors of an SPWS is the feature of variable-frequency operation. Traditional water pumping using grid or diesel is typically configured to operate on constant pump speed, that is, the pump is designed to start and operate at a certain fixed minimum speed (usually 50 Hz or 60 Hz).

As seen in Chapter 2, available solar energy fluctuates throughout the day depending on the irradiation from the sun, thus limiting the operation of a fixed-speed pump. Solar pumping technology has therefore been engineered to overcome this hurdle. This means that solar pumping systems are designed to be able to start even at low frequencies and to adjust the operating frequency according to the available energy from the sun. Consequently, the flow delivered by the pump also fluctuates relative to the speed of the pump, allowing water to be delivered throughout the solar day, albeit at low quantities in the morning and evening when the sun's intensity is low (Figure 3.1).

3.2 SPWS configurations

Different solar pumping configurations are possible based on the power source (solar stand-alone vs hybrid) and water source (submersible vs surface).

SOLAR-POWERED WATER SYSTEM CONFIGURATIONS AND COMPONENTS 29

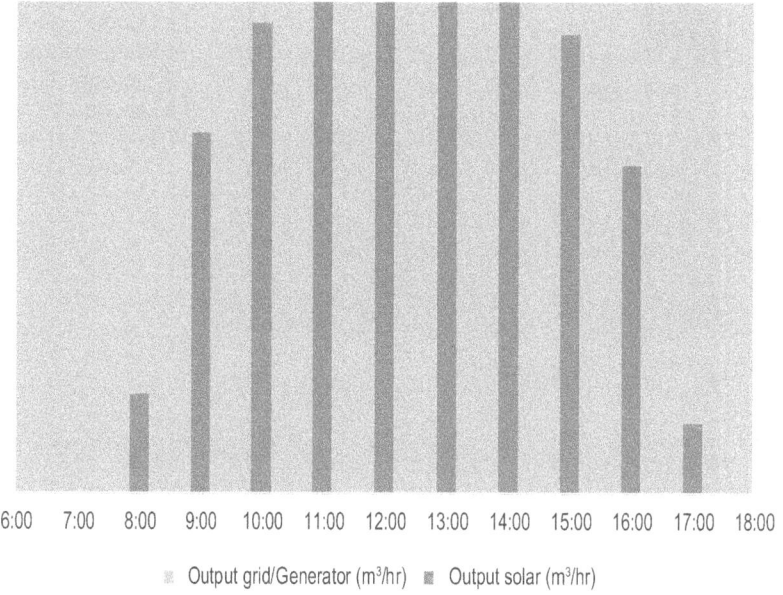

Figure 3.1 Typical flow output profile of a solar-powered pump
Note: Light grey area shows the output of a constant/fixed speed pump running on generator or grid power during the same period

3.2.1 Solar stand-alone vs hybrid configuration

A stand-alone SPWS has solar as the only source of power. It consists of a PV array connected to a pump assembly via controller, as shown in Figure 3.2.

A hybrid SPWS will have solar as the primary source of power with an alternative source of power connected, such as AC power from either a diesel generator or grid supply, for pumping when solar energy is insufficient to run the pump (Figure 3.3). It enables pumping where water demand cannot be met using solar and pumping is required beyond the solar day; or in cases where the water source is constrained and cannot provide enough water over a 4–9 hour solar pumping day; or where water demand is variable with the seasons and a supplementary power source can be switched on as needed to meet the increased demand (intermittent usage). More of this is discussed in section 5.3.8.

Whenever possible and where the solution can meet the demand duty, solar stand-alone should be prioritized over hybrid as it is the most cost-effective with the shortest payback period. Such a decision should always begin with availability of full water-source data, such as a test pumping report, as well as reliable water-requirements data.

30 SOLAR PUMPING FOR WATER SUPPLY

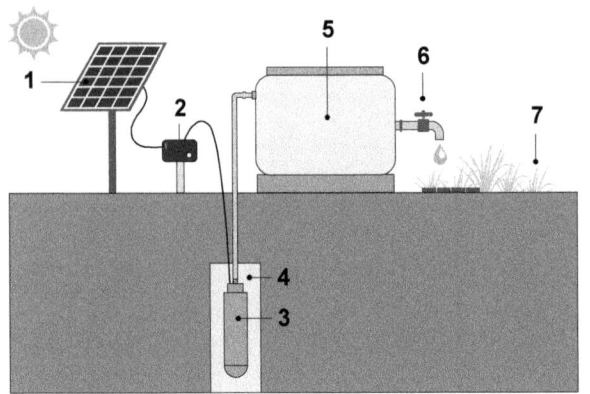

Figure 3.2 Stand-alone solar installation

1 PV modules
2 Controller
3 Pump
4 Water source
5 Storage
6 Tap stand
7 Irrigated crops

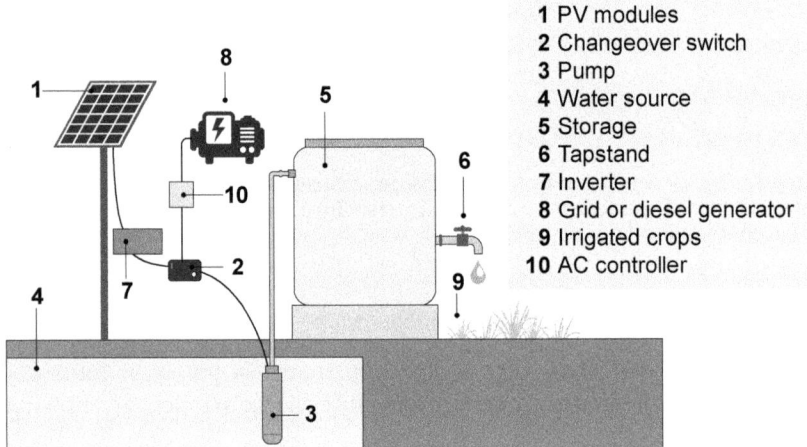

Figure 3.3 Hybrid installation

1 PV modules
2 Changeover switch
3 Pump
4 Water source
5 Storage
6 Tapstand
7 Inverter
8 Grid or diesel generator
9 Irrigated crops
10 AC controller

3.2.2 Submersible vs surface configuration

Submersible configuration has a submersible pump installed completely submerged in the water source (mainly in wells and boreholes). Submersible pumps can also be installed horizontally inside water reservoirs at a minimum depth of 0.5 m, well anchored and fitted with a cooling sleeve (plastic or metallic cylindrical shroud put around the motor) to enable enough cooling to the motor.

Surface configuration has the pump mounted outside and near the water source (e.g. tank, river, dam, lake). They are mounted at ground level with the inlet connected to the water source through a suction pipe and the outlet to the delivery pipe (Figure 3.4).

Both submersible and surface pumps can also be installed floating in a surface water source such as a river, lake, or a dam. This is applicable in

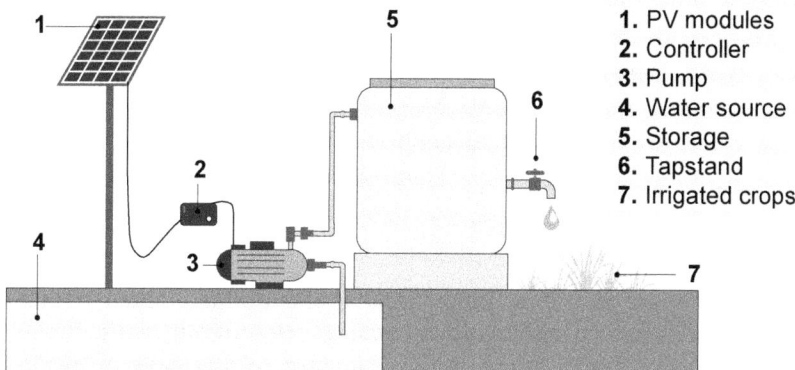

Figure 3.4 Stand-alone surface Installation

situations where it is not possible to install the surface pump close enough to the water source, which would otherwise result in a very high suction head that could cause cavitation problems (see section 3.3.1, 'Surface pumps vs submersible pumps'). Where water levels fluctuate, to mitigate against flooding the surface pump in the rainy season and to manage the suction lift in the dry season, the pump is installed floating on the water source.

The optimal configuration (solar stand-alone or hybrid) is determined based on multiple criteria:

- *Solar resource* – locations where peak sun hours are insufficient to meet demand will require a hybrid configuration for prolonged pumping beyond the solar day.
- *Prevailing weather* – some locations have seasons when the weather is overcast, necessitating a hybrid system which will allow intermittent diesel pumping during prolonged periods of cloud cover.
- *Water demand* – where the water demand exceeds that which solar stand-alone can provide, a hybrid system is necessary.
- *Water source* – a water source that has a limited flow will result in a small pump that will require prolonged pumping beyond daylight hours to meet demand. A surface water source will typically be equipped with a surface pumping system.
- *Economic reasons* – solar systems have a low cost of ownership and a short payback period, making it preferred over hybrid systems.
- *Demographic factors* – contexts where the population is unknown, uncertain, or expected to fluctuate unpredictably should be installed with a hybrid system to cushion against water supply fluctuations and shortages when the population increases.
- *Social aspects* – some communities may have low acceptance and may resist installation of solar, necessitating installation of a hybrid system until there is wide acceptance of the technology.

In addition, some scenarios call for specific contextual considerations, such as:

- Locations where fuel supplies are irregular or difficult to transport to site due to inaccessibility and/or security will adopt solar stand-alone even if there is no economic rationale for solar.
- Critical water supplies which would affect an entire population or bring operations to a halt when the water system is down would require a hybrid configuration to ensure supply continues when one power supply is faulty.
- Where spare parts or technical services are not immediately available, the use of hybrid systems ensures continued pumping while the solar system is being repaired, which can take several days to weeks in remote locations.

Application note

Specific equipment constraints may also influence the choice of configuration in favour of solar, but with a diesel option to meet demand. For example, considering a demand of 600 m^3/day and the largest solar pumping solution available in the country/local market is a 30 kW pump giving 40 m^3/hr. Instead of installing a 55 kW diesel system with a flow of 60 m^3/hr running for 10 hours to meet the demand, the 30 kW solar system can be installed to run on solar for approximately 7 hours (40 m^3/hr x 7 hr = 280 m^3/day), with night pumping for 8 hours on diesel to meet the shortfall of 320 m^3/day. Comparing the fuel consumption of the two options, the 55 kw pump would be connected to a 140 kVA diesel generator consuming 25 litres x 10 hours = 250 litres of diesel per day. With the second option, the 30 kW pump will be connected to a 75 kVA generator consuming 15 litres x 8 hours = 120 litres of diesel. The cost benefit of choosing the second option is obvious. Note if a 55 kW solar pump solution was available there would be no need to pump using diesel, it is the SPWS equipment constraints in this context which necessitates the need for a hybrid system.

3.3 SPWS components

A solar-powered water scheme will have the DC side (entire part of the system that generates and conveys DC power, including the PV generator, DC terminals, DC isolators, DC switches, DC cables, controller, DC pump) and/or the AC side (entire part of the PV system which begins after conversion of DC power to AC power, including AC switches, changeover switch, AC cables, AC pump). All electrical components are rated for the form of power they will be used to convey, so equipment used to convey DC power is rated for DC, and equipment used to convey AC power is rated for AC. They must be designed and used in compliance with the relevant electrical codes.

A solar water pumping system has three major components which constitute on average 60 per cent of the total cost of the scheme, with the balance of system components (DC accessories, cabling, drop pipes, etc.) accounting for the remaining 40 per cent.

- Pump – the pump can be either DC or AC type, centrifugal or helical rotor type, surface or submersible type.
- Controller – DC pumps use DC power directly from the solar modules through a simple control box. For an AC pump, an inverter is required to invert DC power from solar modules to AC power required by the AC pump.
- Solar module – the solar modules generate DC power and supply it to the pump through the control box or inverter.

3.3.1 Water pumps

Pumps are rated in flow (m^3/hr) and head (m) and are connected to an electrical motor that is rated in kilowatts (kW). The motor draws power from the source (solar PV generator in this case) and drives the pump to deliver water.

Solar pumps can be broadly sorted into three categories:

- the motor power type – solar DC vs AC pumps;
- the pump end design: positive displacement vs centrifugal pumps;
- the installation set-up: submersible vs surface pumps.

The pump set should be matched to the water source, the power source, and the application.

Selection of the appropriate pump size is based on the duty point, which is the required flow output and head. Manufacturers' pump performance curves or computer software are used to determine this. An explanation of the selection process is provided in Annex A and Annex B.

A key factor in pump selection is ensuring that the pump duty point is at the best efficiency point (BEP). A common practice is to select the duty point to the right of the BEP on the pump performance curve, so that as the pump wears out the duty point shifts towards the BEP, thereby achieving efficient operation over the life of the system.

Solar DC pumps vs solar AC pumps. Both DC and AC pumps are available for use. The distinguishing feature between DC and AC pumps is in the motor.

A DC pump system is the simplest SPWS configuration and consists of a PV array directly connected to a pump assembly with a DC motor via a DC controller. DC pumps have longer lifespans and are more efficient compared to an equivalent size of AC (up to 90 per cent versus 50–70 per cent for AC) as no power conversion is necessary. These pumps are, however, limited in head and flow and are generally used for lower head, lower volume (i.e. smaller) applications of up to 4 kW power demand. The pump design can be positive displacement or centrifugal type. The motors can be either brushed or brushless (both have permanent magnets).

The brushed motors have brushes that deliver current to the motor windings through commutator contacts, while brushless motors have none of these commutators. Brushed motors have the advantage of being less costly to buy with simple installation as they can be wired directly to DC power through a simple switch without the need of complicated electronics. Yet they are less efficient (75–80 per cent), electrically noisy, and have a maintenance cost due to wearing out of the brushes and commutators.

Brushless motors on the other hand, have a higher efficiency (85–90 per cent), longer lifespan, and are less costly to maintain as they do not require replacement of brushes. They do cost more than brushed motors and have an additional cost of an encoder and a driver to control.

In an AC pump system, the PV array is connected to a pump assembly with an AC motor. The motor is typically a brushless 3-phase induction (asynchronous) motor. These motors have a robust design with standard or enhanced insulation providing long, reliable service, minimum maintenance, and ability to withstand the voltage stresses encountered with most inverter drives. AC motors cannot operate with DC power and require a DC–AC inverter to convert incoming DC supply to power the AC pump. The pump design is commonly centrifugal due to their high flow capabilities.

An AC pump system is used for higher capacity applications that cannot be handled by a DC system.

Positive displacement vs centrifugal pumps. Examples of positive displacement pumps are helical rotor, diaphragm, or piston types. A helical design is used in submersible pumps. It features a rotor cased inside a rubber stator that spins with the motor, creating a vacuum that allows water into the cavity and effectively squeezes water out as it rotates. They deliver water with every rotation, with water output increasing with rotational speed, meaning their efficiency and lift capacity remains high even at low rotational speeds. Consequently they are appropriate for the varying solar radiation levels of solar-direct water pumping. This means higher volumes of water can be pumped per day in variable solar conditions. These pumps all fall within the DC range of solar pumps and therefore have a limited capacity, suitable for high heads and low-flow applications. Heads of up to 450 m are achievable while maintaining high efficiency.

Centrifugal pumps feature one or more impellers inside a chamber (referred to as a pump stage). Water enters the eye of the impeller and as the impeller spins the water is subjected to centrifugal force that pressurizes the water from one stage to another. To achieve high lift, multiple stages are stacked together (multi-stage pumps) with the pressure of the water increasing as it is pushed from one stage to the next. This is the reason high-pressure centrifugal pumps are tall, (submersible pumps can have up to 100 stages with heights of up to 6 m!).

Centrifugal pumps require a minimum speed to start and deliver water. They can achieve flows of up to 250 m^3/hr with efficiency reducing at high heads and low flows. For this reason, positive displacement pumps are used for most systems that require high lift at low volumes. The efficiency of centrifugal pumps deteriorates as the speed varies, whereas positive displacement pumps can operate efficiently over a wide speed range. Positive displacement pumps can also operate at fairly constant flow over a wide pressure range.

Surface pumps vs submersible pumps. Submersible pumps are installed completely submerged in water. They are predominantly used for deep-well pumping. Submersible pumps are coupled to water-cooled or oil-cooled motors and must never be operated without water otherwise they will burn out due to dry running. They are available in both centrifugal and positive displacement designs.

A surface pump is installed outside the water source – it cannot be submerged. A surface pump has an air-cooled motor which should be installed in a well-ventilated location protected from the weather. It is prone to failure if it is submerged or splashed with water. Commonly, surface pumps are installed inside a pumphouse. They are designed to draw water from a depth of 3–7 m above the water-source surface level. If this vertical height (referred to as suction lift) is exceeded, the pump experiences cavitation (as the water enters the inlet of the impeller, low pressure causes it to vaporize, forming bubbles which collapse and erode the impeller as they collapse), eventually leading to pump failure. Surface pumps can be of centrifugal or positive displacement (diaphragm or piston) design.

Surface pumps tend to be more efficient at high-flow pumping and are less expensive than submersible pumps, but more complex to install and operate.

Manufacturer data sheets will state the type and name of pump.

Quality and performance considerations. Pumps are recommended to meet EN 809 and EN 60034-1 or equivalent standards, stainless steel with a minimum grade of AISI 304 or higher.

Centrifugal pumps may be constructed of other materials, such as cast iron and plastic, the choice of which will depend on various factors, such as the quality of water to be pumped. For example, water that has a lot of silt is better pumped with a pump of plastic internal construction, whereas corrosive water or hot water is better pumped with pumps of higher grades of steel. Importantly, the material used for pump construction should be corrosion resistant, permanently lubricated, and maintenance free, as well as able to handle the water temperature.

The pump motor is the piece of equipment most prone to failure and should be constructed with corrosion-resistant material, all stainless-steel exterior construction, stainless-steel shaft, ceramic bearings, NEMA mounting dimensions, hermetically sealed windings, water lubrication, and pressure-equalizing diaphragm, and able to withstand a certain maximum temperature.

The pump set must be of modular design to allow for replacement of individual parts (pump end, pump motor, and electronics) if failure occurs. The pump must have dry-run protection to protect it in the event of low water levels.

3.3.2 Controllers

Solar panels produce DC power and the controller acts as a power conditioner, meaning it improves the quality of power that is delivered to the pump from the PV modules. It acts in one or more ways to deliver a voltage of the proper level and characteristics to enable the pump to function properly. Another important function of the controller is to start the pump slowly (soft start) and adjust its speed according to the pumping load and power available from the solar array. Maximum power point tracking optimally matches the power output from the solar array to the load throughout all conditions.

Solar pump controllers are rated in watts or kilowatts. Selection of the appropriate size to power the pump is based on the allowable input power (minimum and maximum DC voltage and current), controller power output (voltage, current), and the power rating of the pump motor (voltage, current). The controller influences the series/parallel configuration of the modules. Manufacturer data sheets or design software can be used to determine each of these. A full explanation is provided in Annex B.

DC controllers. DC pumps utilize DC solar controllers, which come in two forms, depending on the type of motor used in the pump.

Brushed DC motors can be wired directly to DC power through a simple switch without complicated electronics. A brushed helical pump will have a linear current booster (LCB) that reduces the voltage from the PV array while it boosts the current. This starts the pump motor and prevents it from stalling during low-light conditions. The brushed centrifugal pump is often supplied without an LCB because it starts easily, and its current draw diminishes with speed. An LCB will increase the pump's efficiency during low-sun periods, but the performance gain is relatively small.

Brushless DC motors require an external drive to control the voltage and current applied to the motor, perform the LCB function, and match the pump speed to the available power. The controller varies the motor speed by varying its own frequency. A brushless DC pump is normally sold with a controller that is engineered specifically for it, for example, Lorentz Brushless DC motors have an external driver that is mounted on the surface to control the pump. Controllers for brushless motors sometimes come integrated into the motor, such as the Grundfos SQFlex, and Nastec solar pumps. The disadvantage of integrated controls is that if there is a problem with the electronics in the motor, the entire pump and pipe assembly must be removed from the well and the entire motor assembly replaced. The above-ground controls are versatile and easier to access for repairs and servicing.

Solar DC controller ratings are typically up to 4 kW and the rated power is DC voltage and current.

AC inverters. For AC pumps, a solar inverter is needed to convert DC power output from the PV array to AC power for the pump. The inverter outputs a variable frequency depending on the energy available from the sun, and consequently enables the pump motor to run at variable speed with varying flow output depending on energy availability. It also works to smooth the sinusoidal AC wave form and maintain a constant voltage over varying loads. Ratings of up to 150 kW are available. Most inverters for pumping are of 3-phase AC power output, the reason being that single-phase inverters have little relevance in pumping due to the availability of direct DC-powered solar pumps.

Inverters are now available with power blending or dual power supply capability. The specific requirements of such inverters, such as isolation switches to prevent both powers being present at the same time, are usually provided by the manufacturer. The technical specification gives detailed information on how they should be mounted, protected, and operated.

Quality and performance considerations. A good solar controller must meet EN 61800-1, EN 61800-3, EN 60204-1, or internationally recognized equivalent standards that are indicators of equipment safety and quality. Design and use of a controller must follow IEC 62109-1 and IEC 62109-2.

DC solar controllers and AC inverters should also have the following features for control, monitoring, and protection to make pumping practical and efficient:

- integrated MPPT which tracks the maximum power point of the modules to provide a constant voltage to the pump, as discussed in section 2.10;
- provisions for various control inputs, such as dry-run sensors (to prevent dry running) and level control/pressure switch (high-level/low-level tank automation);
- capability to control the pump system and provide diagnostic indicators to show status;
- simple system health indicators that are user visible for troubleshooting purposes – typically pump status, pump speed, well dry, tank full, amperage, power, voltage, temperature;
- easy to service and unit replaceable by a trained person with modest skills;
- protections for over current, under voltage, over speed, over temperature, reverse polarity, and dry running;
- protection against overload, such as when the pump/pipe becomes clogged with dirt (such occurrences cause an increase in the current consumed, which could lead to motor failure);
- direct solar connection as standard and the ability to add on an optional power backup;
- suitable for outdoor installation (IP54 and higher – sealed, weatherproof, insect proof, rodent proof);
- designed for >10 years lifespan under harsh environmental conditions and have a high efficiency, typically >97 per cent.

Other optional features include data logging of operating parameters which can be recalled for reference and troubleshooting.

3.3.3 Solar modules

To generate usable power, PV cells are connected in series and parallel to produce solar modules with varying current, voltage, and power outputs. The cells are connected in series, grouped, laminated, and packaged between sheets of plastic and glass, forming a PV module as shown in Figure 3.5. The number of cells in a module depends on the application for which it is intended. Modules can have 12, 24, 36, 60 or 72 cells with the power rating of the module increasing with the increase in number of cells. The module has a frame (usually aluminium) that gives it rigidity and allows for ease of handling and installation. Junction boxes, where conductor connections are made to transfer power from the modules to loads, are found on the backs of the PV modules.

Module efficiencies have increased over the years with commercially available modules that are 15–22 per cent efficient, and research laboratory cells that demonstrate efficiencies above 40 per cent.

Types of solar module. There are four main types of photovoltaic module: mono-silicon (mono-Si), poly-silicon (poly-Si), amorphous silicon (a-Si), and thin-film (Figure 3.6).

Mono-silicon (monocrystalline, mono-Si) contains a higher purity silicon (a single continuous crystal structure), resulting in the highest efficiency (15–20 per cent). It is consequently the most space efficient and produces more power for an equivalent surface area, but is also the most expensive. Mono-Si cells are easily recognizable by an external even colouring and uniform look with rounded edges. They have guarantees of up to 25 years.

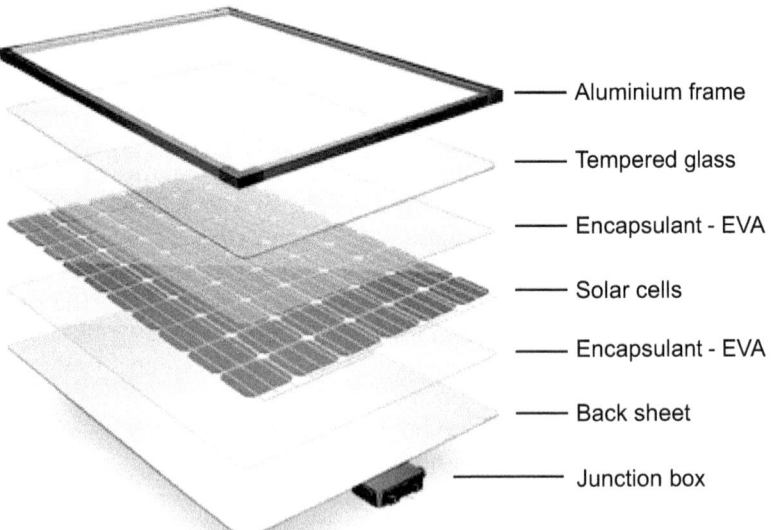

Figure 3.5 Illustration of a solar module construction
Source: Trina solar

Poly-silicon (multicrystalline, poly-Si) is made from various silicon crystals formed from an ingot and is simpler to produce and therefore slightly cheaper than mono-Si, however it tends to have lower output efficiency (13–18 per cent). Poly-Si cells look perfectly rectangular without rounded edges. They have guarantees of up to 20 years.

Amorphous silicon (a-Si) lacks a geometric cell structure. Amorphous modules do not have the ordered pattern characteristic of crystals as in the case of crystalline silicon. Commercial modules typically have conversion efficiencies from 5 per cent to 10 per cent. Most product guarantees are for 10 years, depending on the manufacturer.

Thin-film photovoltaic cells (TFPV) use a semiconductor such as cadmium telluride (CdTe) or copper indium gallium selenide (CIGS or CIS) that is applied as a thin film or layer onto a substrate (such as glass, plastic, metal, ceramics). The modules have a solid black appearance. The thin layer allows thin film modules to be flexible, lightweight, and portable. They are cheaper to produce but much less space efficient, therefore needing a larger support structure and more cabling. A monocrystalline module of equivalent area can typically produce four times the amount of electricity produced by a thin-film module.

The energy consumption during the production process in a-Si and TFPV is lower than that needed for c-Si (mono-Si and poly-Si) modules. The influence of the temperature on the output power of TFPV is also the lowest of all PV technologies. Since 2010 some thin-film technologies (CIS, CIGS, and CdTe) have been used in large-scale grid-connected PV systems, with the same warranties as the c-Si modules. Smaller efficiencies (10–18 per cent) are compensated for with a reduction in the price. The main drawbacks of the low efficiency are the larger dimensions of the PV plant, cost of the structure, and assembly time.

Thin-film technologies are evolving and they have enormous potential in the future as efficiency improves and costs reduce. One thing to note is that some thin-film modules, such as cadmium telluride, use toxic cadmium which, if not recycled, poses an environmental concern.

The choice of which module to use usually depends on what is available in the local market (all quality considerations taken into account), cost, and space efficiency. The recommended modules for water pumping applications are mono-Si and poly-Si because of their higher performance, space, and cost efficiencies.

Module characteristics. After testing PV modules at STC (section 2.7), manufacturers will rate the modules and label them. It is important to select modules that are clearly labelled and permanently marked with a data plate. The plate will contain the following information: manufacturer's name and physical address, type/model number, the watt-peak power rating, open-circuit voltage, short-circuit current, voltage and current at maximum power point, tolerance, temperature coefficient, country of manufacture, and certification

Figure 3.6 Different types of solar modules (clockwise from top left: mono-Si, poly-Si, a-Si and TFPV)

Table 3.2 Example of module characteristics

Characteristic	Term	Unit	Value
Peak power	P_{max}	W_p	250
Power tolerance		%	0 to +5
Maximum power point current	I_{mp}	A	8.23
Maximum power point voltage	V_{mp}	V	30.4
Short circuit current	I_{sc}	A	8.81
Open circuit voltage	V_{oc}	V	37.6
Temperature coefficient for P_{max}		%/°C	−0.42
Temperature coefficient for V_{oc}		%/°C	−0.34
Temperature coefficient for I_{sc}		%/°C	0.06
Maximum system voltage		VDC	1,000
Module efficiency		%	17.12
All values at STC (AM = 1.5, E = 1,000 W/m², cell temperature = 25°C)			

(e.g. UL listing, IEC 61215, ISO certification). Table 3.2 gives characteristics for a typical 250 W module.

These characteristics are important for designing the required solar generator size and determining the module arrangement (series and parallel), according to the voltage, current, and power limits of the pump and controller (see section 5.3.5 and Annex B).

The cable terminals at the back of the module must also be clearly labelled with positive and negative markings. The terminals come with male (positive) and female (negative) quick connectors for easy, fool-proof interconnection.

The number of modules required to power a pumping system, together with their arrangement in series/parallel, is based on the power requirement of the pump and is influenced by the controller characteristics (explained in Annex B). The modules should be arranged to provide enough current, voltage, and power to the pump, taking into consideration the envisaged losses discussed in Chapter 4 (cloud, dirt, degradation, temperature, etc.).

It is critical to get the series/parallel wiring correct. This is a common problem that can lead to suboptimal pump performance or damage to equipment. The completely wired PV generator should always be checked against the design specification before connecting to the controller/inverter and the pump.

Quality and performance considerations. The key criteria for checking the quality of solar modules is the certification. Crystalline PV modules must be approved to IEC/EN 61215 and 61730, while thin-film modules must be approved to IEC/EN 61646, or all types must be UL 1703 certified and listed, as indicators of quality and adherence to safety standards.

Optional standards apply depending on the actual conditions the modules will be installed in. For example, IEC/EN 61701 is required for modules that will be used in coastal areas; it is an indicator that the module will be able to withstand the salty mist conditions of coastal installations.

IEC/EN 61215 involves the examination of all parameters which are responsible for the ageing of PV modules and describes the various qualification tests based on the artificial load of the materials. As the modules cannot be tested over a period of 25 years, accelerated stress is performed which involves radiation testing, thermal testing, and mechanical testing. IEC 61730 Part I and II is a testing for safety qualification.

There are many brands of PV modules worldwide, some of the leading manufacturers being Canadian Solar, Trina Solar, First Solar, Jinko Solar, JA Solar, Sunpower, Yingli Green Energy, Sharp Solar, Renesola, Hanwha SolarOne, Kyocera, and SolarWorld. While all these manufacturers have their modules properly approved to the above-mentioned manufacturing certifications the industry is competitive and dynamic. Being the largest manufacturer does not necessarily guarantee the highest quality module and some of the smaller manufacturers may also offer premium products.

However, a company's proven ability to produce and sell a large number of solar modules is testimony to its brand credibility.

In general, a good brand should have a quality manufacturing process, a good reputation, be free of defects in the manufacturing process, and have a replacement warranty (see more on the quality of solar modules in section 10.6).

3.4 SPWS balance-of-system components

Balance of system (BoS) components in an SPWS refers to all the components required for the sound installation and operation of the PV system other than the photovoltaic modules, pump, and solar controller. As the name suggests, it provides a 'balance' between the solar modules, controller, and pump. In a PV pumping system, BoS components include disconnect switches, PV combiners, mounting structures, enclosures, wiring to connect different hardware components, sensors to automate or protect the system, meters to monitor the performance and status of the system, flow meters, and so on.

3.4.1 Module mounting structure

The module mounting structure is a critical part of any solar installation. It holds the solar modules and provides safety and security for them. Different structure configurations exist, namely ground mount, pole mount, roof mount, and ground screws, as discussed in section 6.2.5.

A structure can be made of stainless steel, galvanized mild steel (with stainless-steel hooks/bolts), or aluminium. The configuration and material to deploy is determined by factors such as environmental conditions, system size, area topography, land availability, and security concerns.

The mounting structure should be designed to bear the weight of the solar array and to withstand wind loading, snow loading, storms, and/or earthquakes as per local conditions. The footing of the structure should be made of reinforced concrete according to the soil type, with adequate bracing provided to prevent swaying. The structure can be designed using computer-based software to come up with the right sizes of reinforcement, poles, runnings, struts, and ties, in compliance with structural design standards and local building codes.

3.4.2 DC disconnection/isolation switches

DC electricity is generated though the photovoltaic effect of solar PV modules. If PV modules are exposed to sunlight, they will produce electricity. Large amounts of power (voltage and current) will be present in a PV array during daylight hours and caution should be taken when handling the PV array. The power present in the PV array should be isolated before attempting to work on the PV pumping system. A DC disconnection switch (also known as a disconnect switch or isolator switch) provides for this isolation and for subsequent isolation during future maintenance. Some DC disconnect switches also allow for safe and professional paralleling of multiple strings coming from the PV generator. The DC isolator switch is installed between the PV array and the solar controller.

The DC isolation/disconnection switch should be rated for the maximum possible DC current (short-circuit current) in the system and the maximum possible DC voltage (open-circuit voltage). It should also be double pole to effectively isolate both negative and positive inputs from the PV array.

3.4.3 Surge protector

A surge protector (or arrestor) provides enhanced protection for the controller from electrical surges, such as indirect lightning, current surges, and voltage spikes. It is installed on the DC input line between the PV generator and controller, and close to the pump controller.

Lightning arrestors can also be installed in lightning-prone areas to arrest surges due to lightning. A surge arrestor requires a reliable ground connection to operate, as discussed in section 6.3.

3.4.4 Cables

Electrical cables are the carriers of electrical power to the pump. They must therefore be able to deliver enough power to the pump for both the sound functioning and safety of the pumping system. Cable selection should be considered both on the DC side and the AC side.

Two types of insulated multiconductor cable are used in pumping applications: submersible pump cable and underground cable (commonly referred to as UG cable). The submersible pump cable runs from the pump up to the surface of the borehole (wellhead) in submersible applications. It is designed to safely carry electrical loads under water as long as it is sized properly. UG cable, on the other hand, is used between the wellhead and the pump controller. Typically, UG cable is armoured and can be buried in the ground. If it is not armoured it should be run in an electrical conduit. Submersible pump cable can still be used between the wellhead and the controller so long as it is run in an electrical conduit.

A pump cable will usually have the following parts, as shown in Figure 3.7:

- Conductors – these are the copper wires that conduct the electricity.
- Insulation – this is the plastic or rubber material covering the copper conductors to keep the conductors from shorting between themselves or to ground.
- Jacket – this is a rubber or PVC material covering the insulated conductors, protecting them from abrasion.

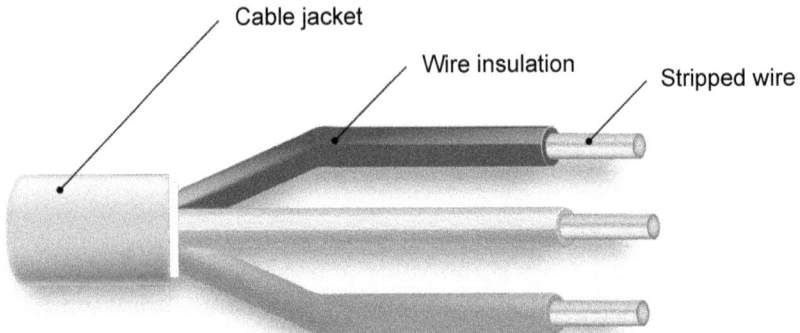

Figure 3.7 Parts of a cable
Source: Wikipedia <https://creativecommons.org/licenses/by-sa/3.0/deed.en>

The correct cross-sectional dimension (or thickness) of the cable is critical for sound operation and long life of the pumping system. Voltage drop or loss occurs when too much current passes through a very small (undersized) conductor. Voltage loss can be reduced by using thicker wires (lower resistance conductors), reducing wire length, and/or increasing voltage (often not practical) (see also section 2.9).

The maximum allowable voltage drop between the source and the load should be 1–3 per cent of the source voltage. For DC power this refers to the maximum power point (MPP) of the string. A suitable cable cross-section should be selected to ensure the voltage loss does not exceed the permissible drop. A cable of insufficient cross-section will convert the current flow into heat, leading to overheating which could cause a fire or progressive failure of the system.

In summary, when selecting electrical cable consider:

- suitability of the cable to the environment (e.g UV strength, ambient temperature, water temperature). All exposed cables should be approved for outdoor use or installed in an electrical conduit;
- suitability of the cable to the application (water-resistant cables for submerged installations, armoured cables for surface installation);
- current carrying capacity of the cable. This is dictated by the cable size or cross-sectional area of the cable;
- allowable voltage drops across the cable. Voltage drop increases with smaller cable thickness and longer cable lengths. Excessively long cables should be avoided where possible to minimize the risk of voltage drop;
- characteristics of the load (voltage and current).

A guide on how to select a suitable cable size is provided in Annex D.

3.4.5 Sine-wave filter

As the external controller/inverter adjusts the incoming voltage for AC motors, their sine wave is altered and in turn taxes the motor lifetime and increases noise. Sine-wave filters have a high degree of filtering and are used for reducing the voltage stress on the motor windings and the stress on the motor insulation system as well as for decreasing acoustic noise from a frequency-controlled motor. Motor losses are reduced since a sine-wave filter converts the output pulses of the frequency converter into a sine-wave shape. The result is a sine-wave-shaped current and reduced motor noise. The sine-wave filter is installed between the inverter and the pump.

In addition to peak voltage effects, total motor cable length should be considered in the context of instantaneous current peaks, which can cause stress on the motor. Filters may be used to extend the maximum cable length according to the technical specifications of the particular inverter type. For example, Grundfos sine-wave filters extend the maximum cable length to 300 m.

3.4.6 Other components

Combiner box. This can be used to wire incoming strings in parallel or to combine two or more PV disconnect switches in parallel.

Dry-run sensor. Dry-run protection is mandatory for any pumping installation. For submersible pumps this is provided in the form of a well probe that is connected to the low-level sensor input of the controller. For a surface pump it comes in the form of a low-level float switch or a water sensor. The purpose of dry-run protection is to stop the pump in the event of low or no water in the source. It prevents dry running which can lead to pump failure.

Float switch. This is a mechanical device used for automation of the pumping system to a tank. It works by stopping the pump when the tank is full and restarting it when the tank water level drops to a pre-set level (see more in section 7.2).

Pressure switch/sensor. This is installed on the delivery pipe to automate delivery to the tank where long delivery pipes are involved and a float switch cannot be used. It works by sensing the pressure build-up of water in the pipe and switching the pump on and off according to the pre-set cut-in and cut-out system pressure. It is installed together with a ball valve at the tank (see more in section 7.2).

Irradiation sensor. Some equipment manufacturers, such as Lorentz, provide this sensor to cut off the pump when the sun irradiation drops below a pre-set level. It plays the important role of avoiding pump idling where the pump runs without delivering any flow, as this would cause the pump to wear out due to lack of internal lubrication and will eventually cause failure.

Water meter. Installed on the delivery pipe to measure and display the water flow output from the pump, the water meter is sized according to the pump flow and the delivery pipe size.

46 SOLAR PUMPING FOR WATER SUPPLY

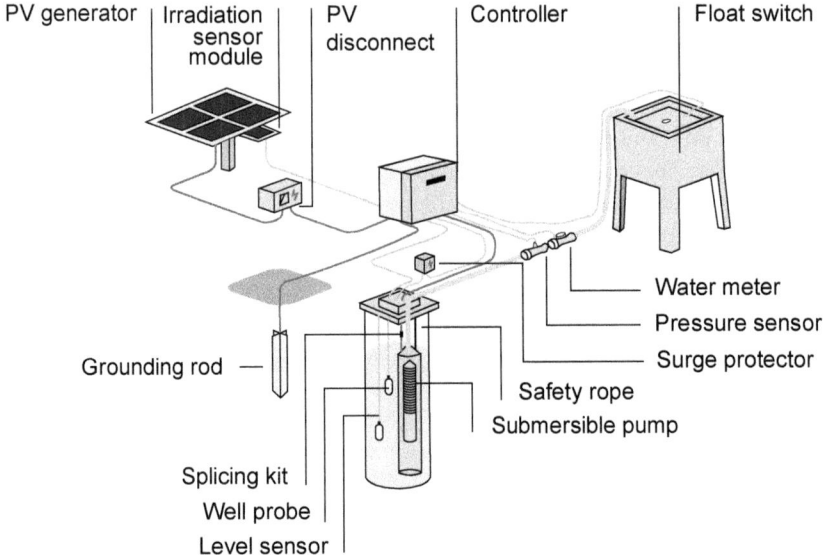

Figure 3.8 The components of a solar water pumping system

Splicing kit. The connection between the motor tail cable and the submersible drop cable must be made watertight as it can be the weakest point on the pumping system. The splicing kit needs to be appropriate for the cable size and should be done by an experienced technician.

Chlorine doser. These are used to inject appropriate amounts of chlorine into the delivery line to sanitize the water (see more in section 7.1).

Advanced controls. Some brands of inverter (e.g. Lorentz) feature advanced controls for digital remote/contactless monitoring of certain parameters that can be collected for long-term monitoring such as:

- water flowrate/output, through the connection of a pulse cable from the water meter to the digital input of the controller;
- water level, through the installation of a liquid level sensor in the well to determine the actual level of water in the well at any given time;
- liquid pressure, through installation of a liquid pressure sensor in the pipe to measure pressure relative to the atmosphere.

Figure 3.8 is an example of a solar water pumping system showing balance of system components.

3.5 SPWS equipment manufacturers

Solar-powered pumps, controllers, inverters, and accessories are available from various suppliers globally. As discussed in Chapter 10, regard should be given to high-quality products supplied by certified suppliers/manufacturers.

Suppliers/manufacturers should be asked to provide certifications that demonstrate the quality and safety of the equipment. Some manufacturers of solar pumping equipment are:

- Lorentz – German technology with a factory in China. Reliable and high-quality range of solar pumps and inverters from 0.15 kW to 75 kW of pump motor power (www.lorentz.com);
- Grundfos – Danish company with multiple manufacturing locations around the world. Has a reliable solar pumping range of up to 37 kW pump motor power (www.grundfos.com);
- Well Pumps – Belgian company with solar range up to 110 kW (https://wellpumps.eu/en/homepage);
- Solartech –inverters of up to 150 kW manufactured in China and distributed through regional suppliers, such as Davis & Shirtliff Ltd (https://www.davisandshirtliff.com) and Solargen Technologies (https://solargentechnologies.com) in Africa;
- ABB – drives for solar pumps with a range of 0.37 kW to 45 kW (https://new.abb.com/drives/low-voltage-ac/machinery/ABB-solar-pump-drives);
- Franklin – solar pumps and drives from 0.55 kW to 37 kW (https://solar.franklin-electric.com/products/high-efficiency/6-inch-high-efficiency-solar-system/)
- Fuji – solar drives (https://www.fujielectric-europe.com/en/drives_automation/products/solutions/frenic_ace_for_solar_pumping).

3.6 Importance of quality considerations in SPWSs

Solar pumping technology is a robust and reliable technology that can provide many years of reliable water supply. Evidence (UNICEF, 2016; GLOSWI, 2016) shows that where failure (non-functionality or breakdown) of SPWSs has occurred, it is mainly because of pipe and pump-related issues, such as motor failure, broken pipes, pipe bursts, or wiring – issues that are also present with traditional grid or diesel-powered pumping systems and which can be averted at the design and procurement stage. This is only possible if the following conditions are met. First, the design is based on accurate and reliable data, matching the right equipment to the water source.

Second, high-quality components with genuine certifications should be selected. During the design stage the proposed equipment should be fully specified in terms of material of construction, protection rating, and cable insulation gauge so it matches the operating conditions. The quality of SPWSs is discussed in Chapter 10.

Third, careful attention should be given to the entire implementation process, including selection of the intake point in the river, the borehole siting, full borehole development, the casing of the borehole, the dimensions of the equipment, and the operating conditions of the equipment, such as water temperature, ambient temperature, and snow.

Figure 3.9 Worn-out conduit due to high temperatures in Iraq

The selection of good BoS components is as important as the selection of pump, controller, and PV modules. Low-quality BoS components are often responsible for many avoidable maintenance problems for SPWSs in camps and remote villages which can lead to premature failure and non-functionality of the whole system. The goal of any SPWS design is to match its operational life to that of the PV modules, that is, a minimum of 25 years. BoS components should therefore be selected with a long operational life in mind.

Figure 3.9 shows what can happen to a piece of equipment poorly matched to its operating environment. Even the smallest piece of equipment (such as a cable gland or conduit), if the wrong size, of poor quality, and not matched to the operating conditions, can be damaged, leading to premature failure and disuse of the system.

CHAPTER 4
Energy losses in solar photovoltaic energy production

The amount of electricity produced by solar panels and therefore the amount of water pumped, changes during the day and over the year depending on a number of factors. Some of the factors that induce energy losses in the system can sometimes be minimized by designers (e.g. shadows on the solar modules), while others are due to the context and the components used (e.g. losses in cables) and can only be taken into account when sizing a solar pumping scheme. The origin and effect of each of these factors are explained in this chapter, together with the calculation of the performance ratio of a solar pumping scheme as the overall indicator of the efficiency of the system.

Keywords: cell temperature losses, soiling, module shadowing, solar mismatching, angular and spectral reflectance, solar performance ratio.

4.1 Calculating energy losses

In standard test conditions (STC), the energy generated by a PV module can be determined by the product of its power rating multiplied by the time of exposure to sunlight. However, in real field conditions there will always be energy losses that PV modules and the wider solar PV pumping system will experience that need to be factored in. Understanding these losses and how to minimize them will be of paramount importance in ensuring that solar pumping systems provide the expected amount of water.

For any type of PV system, the energy generated ($E_{generated}$), or energy available for consumption (e.g. to power water pumps), can be estimated by the following expression:

$$E_{generated} = P_{pk} \cdot PSH \cdot PR$$

where

- P_{pk} is the peak power of the PV field, obtained as the product of the PV module's power rating and the number of modules used in the construction of the PV field (e.g. 10 modules of rated power at 300 Wp will have a peak power of 300 × 10 = 3,000 W_p);
- *PSH* (or peak sun hours) is equal to the equivalent number of hours per day when solar irradiance averages 1,000 W/m²;
- *PR* (or performance ratio) is defined as the ratio between the generated energy and the theoretical energy that would be generated by the PV field if the modules converted the irradiation received into useful energy according to their rated peak power.

In other words, PR express the reduction of solar energy generation due to different losses in the system including: in the PV modules, due to high PV module cell temperature, low irradiance efficiency reduction, power converter efficiency, power converter downtime, wiring, shading, and soiling, among others. The energy loss factors in a PV system are diverse and sometimes difficult to quantify.

Losses can be classified according to what they are related to:

- sun irradiance: soiling (dust and dirt on PV modules), shading, light incident angle, air pollution, orientation and tilt, snow, and others;
- PV module: cell temperature, power tolerance, low irradiance efficiency, mismatch, module quality, light-induced degradation, ageing and other degradation factors, and module mounting conditions are the main ones;
- power conversion and balance of system: voltage drops in the wiring, power inverter efficiency, errors in tracking the maximum power point, protections, downtime periods for maintenance, breakdowns or malfunctions, and power curtailment are the main ones.

Most of these loss factors are estimated for the usual conditions found in PV systems, being corrected and refined as new developments appear in the market and experimental results are published worldwide. Others, like the losses in the wiring or due to the PV module temperature, are estimated by formulas, taking into account technical characteristics of the components and specific conditions where the PV plant is located.

The closer the PR of a PV system is to 1, the more efficient the PV system is and, therefore, fewer PV modules will be needed to meet the required water needs. In order to estimate the PR of the system, estimates of the losses due to each of the factors mentioned above are needed. In a well-designed and mounted PV system, typical losses are as estimated in Table 4.1.

Table 4.1 Table of estimated losses as a percentage of total energy produced

Losses due to	Estimated losses (as % of total energy produced)	Losses due to	Estimated losses (as % of total energy produced)
Module temperature	8–15%	Tolerance	0–5%
Wiring	1–3%	Mismatching[1]	1–2%
Soiling[2] (dust, dirt on modules)	2–15%	Low irradiance	1–4%
Shading[3]	0–2%	Light-induced degradation	3–20%
Reflectance	2–6%	Power converter	1–5%
Module orientation	0–2%	Availability	1–3%

[1] Can be much higher if PV modules of different power ratings are connected
[2] Can be much higher in dusty environments if modules are not cleaned regularly
[3] Can be higher if modules are regularly shadowed

While different solar design software packages take into account some or most of the losses explained in this chapter, it will be of paramount importance that designers and field practitioners understand the different factors that affect the efficiency of a solar PV pumping system so they can act on them.

The following sections describe the most common loss factors, starting with the losses that can be calculated with precision through mathematical formulas: losses due to temperature and losses in the wiring. Recommendations to minimize losses whenever possible are given at the end of each section.

4.2 Cell temperature energy losses

The peak power of the PV modules is rated at STC, with PV cell temperature at 25 °C. However, in real field conditions, PV cells (or modules) easily reach temperatures over 25 °C when exposed to sunlight. The hotter a PV module gets, the less voltage is produced and therefore power output is also reduced. The more the PV module temperature exceeds 25 °C, the higher the losses due to temperature.

As can be seen in Figure 4.1, the difference between the PV module and ambient temperature can reach 20 °C during the central hours of the day, and even higher depending on the PV module characteristics and location.

Figure 4.2 shows the irradiance (in W/m², in the right axis) and the PV cell efficiency (in per cent, in the left axis). There can be seen a small decrease in the cell efficiency at midday due to the high level of irradiance, which produces an increase in the cell temperature and therefore more losses.

The low efficiency at the beginning and end of the day is related to the optical losses and the low irradiance efficiency, detailed in section 4.5.4.

The power losses due to a module temperature different from the standard test conditions (25 °C in STC) depend on the cell type and encapsulation, wind, and the type of installation carried out. They are quantified by the term (L_{temp}). Using the temperature coefficient (g in 1/°C or 1/K) given by the manufacturer for each PV module and the PV cell temperature (T_{cell}), L_{temp} can be approximately determined by the following expression:

$$L_{temp} = g(T_{cell} - 25)$$

The value of g determines the dependence of the PV output power with T_{cell}. Due to negative sign of g, the output power of the PV module decreases as the T_{cell} increases. The cell temperature can be estimated for a known ambient temperature (T_{amb}) and for a known sun irradiance (E, measured in W/m² with an irradiance meter) using the following expression:

$$T_{cell} = T_{amb} + (NOCT - 2) \cdot \frac{E}{800}$$

52 SOLAR PUMPING FOR WATER SUPPLY

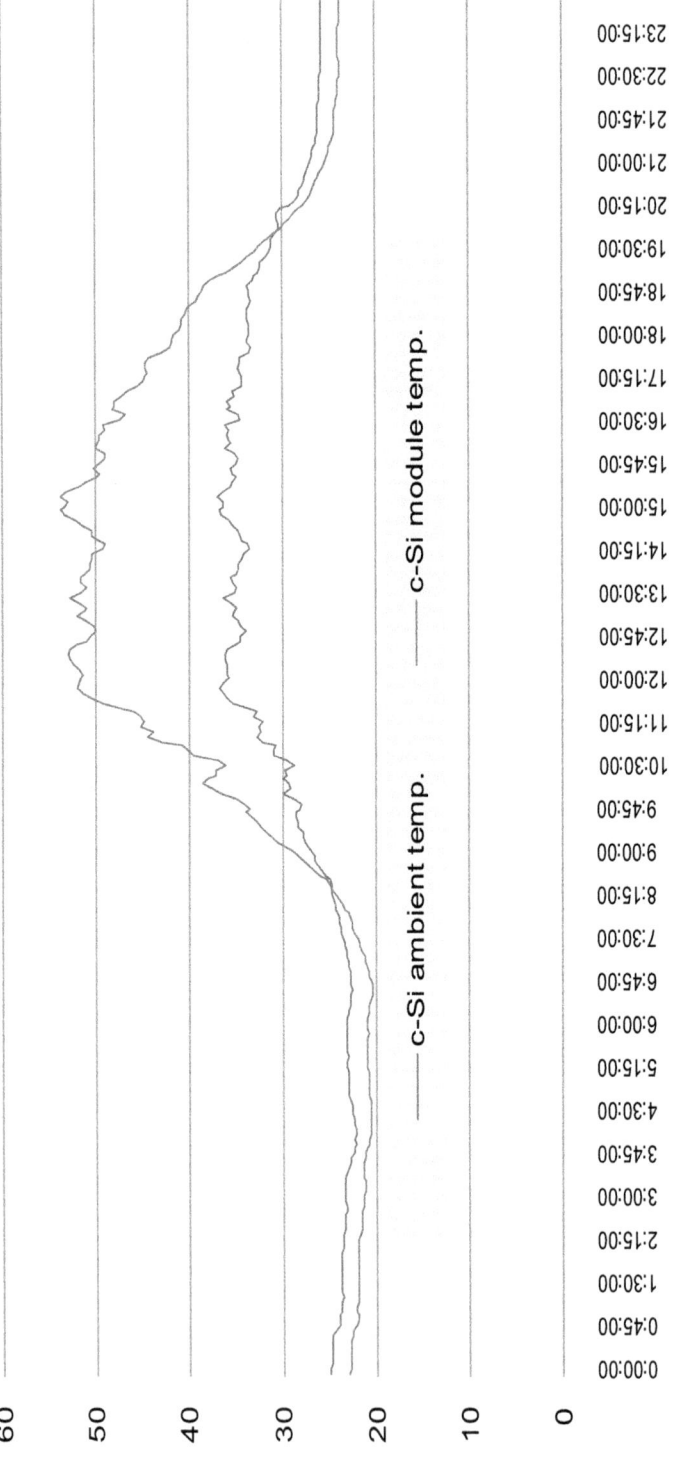

Figure 4.1 Temperature difference for a c-Si plant with a fixed-tilt angle of 30° in Valencia, Spain
Source: Polytechnic University of Valence, Spain

ENERGY LOSSES IN SOLAR PHOTOVOLTAIC ENERGY PRODUCTION 53

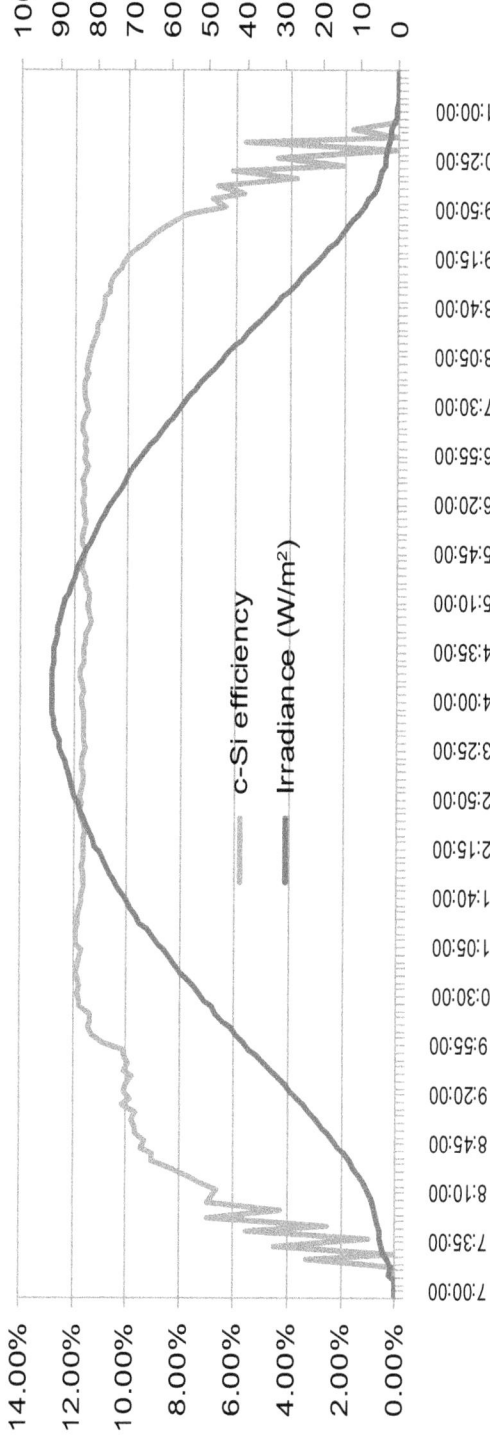

Figure 4.2 Effect of sun irradiance in efficiency of c-Si PV modules
Source: Polytechnic University of Valence, Spain

Table 4.2 NOCT, g, and efficiency factors for different PV module technologies

Technology	c-Si (silicon-based modules)		Thin-film technology			CISG	HIT
	s-Si (monocry- stalline silicon)	p-Si (polycry- stalline silicon)	a-Si/μc-Si (amorphus silicon)	CdTe (cadmium telluride)	CIS (copper indium selenide)		
NOTC (°C)	41 (±3)	41 (±3)	45	45	47	42	44
g (%/K)	−0.37	−0.38	−0.35	−0.32	−0.33	−0.3	−0.26
Module efficiency (%)	20.4	17.6	9.8	10.6	15.1	16	20.3

Manufacturers of PV modules state in their data sheets the nominal operating temperature of a cell (NOTC), which is defined as the temperature that is reached by photovoltaic cells when the module is subjected to an irradiance of 800 W/m² with a spectral distribution AM of 1.5 G, ambient temperature of 20°C, and a wind speed of 1 m/s. Some typical values for the NOTC and g are detailed in Table 4.2 for some commercial PV modules with different PV technologies.

Application note

Considering a monocrystalline c-Si module with $g = -0.45\%/K$ and NOTC = 45°C on a summer day with a maximum ambient temperature equal to 20°C and $E = 1{,}000$ W/m², the estimated T_{cell} is equal to:

$$T_{cell} = 23 + (45 - 20) \cdot \frac{1000}{800} = 54.25 \text{ °C}$$

And the corresponding L_{temp} value is equal to:

$$L_{temp} = g(T_{cell} - 25) = -0.45\ \%/_K \cdot (54.25 - 25) = 13.16\%$$

The measurement of the ambient temperature should be carried out with a thermometer, in the shade, in an area close to the photovoltaic modules.

Due to the variations of the temperature and irradiance during the day and through the year for the specific PV system location, the value of L_{temp} is commonly calculated with the average daytime temperature of the period under study. Most solar design software from the most popular pump manufacturers will estimate this loss by considering ambient temperature data for every month of the year, making the estimation closer to reality. Annual L_{temp} losses between 8 per cent and 15 per cent are common for monocrystalline c-Si modules, depending on the temperature characteristics of the PV system location.

4.3 Wiring energy losses

PV systems convert sunlight into electrical energy that flows as electric current through wires that have an electric potential between terminals. As explained in section 2.9, the flow of a current (I) through a wire produces

> **Box 4.1 Minimizing losses due to temperature**
>
> Installations done in places with good wind levels can have less L_{temp} losses due to the better cooling of the PV modules. Similarly, installing spacers to allow some 15–20 mm of space between PV modules (see photo below) will allow better air circulation, cooling down the PV plant. In roof-mounted PV modules, allowing separation of 10 cm between PV modules and the roof would allow air flow and reduce thermal losses.
>
>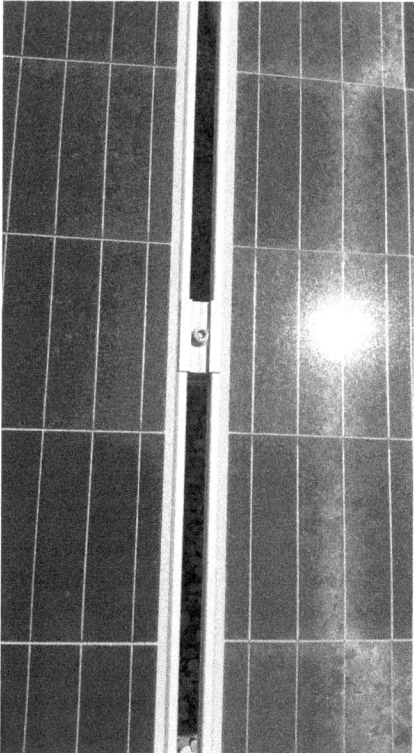
>
> **Figure 4.3** Spacers between modules to increase cooling

a voltage drop (ΔV_{wire}) between the wire terminals due to the resistance of the wire (R_{wire}).

Voltage drop is the reduction in electrical current that occurs as electricity travels through wires. The international standard recommends that voltage drop in the wires connecting solar arrays and pumps not exceed 3 per cent, but it is acceptable to exceed this slightly in order to site the array in a secure location.

This voltage drop causes a power loss (ΔP_{wire}) along all the wire that overheats the conductor. Losses due to wiring are represented by the term L_{wires}, where

$$L_{wires} = \Delta P_{wire} \div P_{pk}$$

with P_{pk} being the peak power of the PV field.

56 SOLAR PUMPING FOR WATER SUPPLY

The power loss in a wire is calculated as the product of the current and the voltage drop:

$$\Delta P_{wire} = \Delta V_{wire} I$$

Wiring losses in PV systems are classified as:

- DC wiring losses, due to cables used to connect the DC components of the system – modules, combiner boxes, DC protections, monitoring systems, DC pumps, etc.;
- AC wiring losses, due to cables used to connect the AC components of the system – output circuit of inverters, AC pumps, protections, etc.

Wiring losses (for copper cables) are calculated with the following expression:

$$\Delta P_{wire} = R_{wire} \cdot I^2 = \rho \cdot \frac{l_{wire}}{S_{wire}} \cdot I^2 = \frac{1}{\gamma} \cdot \frac{l_{wire}}{S_{wire}} \cdot I^2$$

Where l_{wire} is the total wire length in metres (equal to twice the length of the cable for DC cables), S_{wire} is the wire cross-section in mm², ρ is the resistivity of the conductor in Ω mm²/m and γ is the conductivity of the cooper in m/Ω mm². The value of ρ (or γ) must be calculated for the operating temperature of the wire. As an initial recommendation, a wire temperature of 40 °C must be considered for any conductor in a PV system due to the overheating produced by the power loss in the wire. Wires exposed to sunlight will reach greater temperatures which, in some cases, can be near 90 °C (the maximum temperature supported by wires can get to 120 °C for some wire types). Table 4.3 gives the values of the cooper resistivity and conductivity for several wire temperatures.

The typical average value for L_{wires} is in the range of 1 to 3 per cent, depending on the national codes and regulations applicable to each country. Most solar design software from the most popular pump manufacturers will factor in this loss in their proposed designs.

4.4 Sun irradiance energy losses

Sun irradiance is defined as the solar power incident on a surface. There are several sources of losses related to sun irradiance reaching the PV surface: dirt and dust on the modules (also referred to as soiling); near or far shadows; orientation and tilt of the PV module's surface; level of air pollution in the

Table 4.3 Variation of resistivity and conductivity with temperature

Temperature	20 °C	40 °C	50 °C	60 °C	70 °C	80 °C	90 °C
Resistivity (Ω mm²/m)	0.01786	0.01926	0.01996	0.02066	0.02136	0.02206	0.02276
Conductivity (m/Ω mm²)	55.9910	51.9205	50.0994	48.4017	46.8153	45.3295	43.9352

> **Box 4.2 Minimizing wiring losses**
>
> To minimize L_{wires} it is important that, as far as possible, the cables supporting the highest current are the shortest possible in length; placing the PV modules as close as possible to the water pump will ensure that. Due to the large variations in the technical conditions of each pumping installation it is not possible to determine whether it is best to minimize the length of the cable on the DC side (PV modules to controller) or on the AC side (inverter to pump, in case the pump is AC). The use of a spreadsheet to evaluate the different options will allow a decision in each case as to which one provides the least total losses in wiring. However, minimizing the length of the cables on both sides of the inverter, by placing the inverter and the PV modules as close to the water pump as possible, will bring down power losses due to wiring. In addition, protecting the cables from direct sun exposure will reduce their resistivity and therefore the wiring losses.

surroundings of the PV installation; light incident angle; and weather characteristics of the location (snow, atmospheric humidity, rain).

Specific weather conditions of the region where the PV plant is installed can produce a reduction of the irradiance reaching the PV surface. Some of these factors can have a seasonal pattern, so the effects vary throughout the year:

- The presence of snow in winter months will cause the PV system to disconnect until the modules are cleared or the snow disappears. The presence of the snow is related to days with low availability of sun hours, so the impact on energy production is minimal. These losses are included in the availability losses ($L_{availability}$) together with other terms that are described later.
- Rain over PV installations can naturally clean the PV modules, reducing losses due to soiling. At the same time, in cities and industrial regions, the air pollution will be reduced, so more direct radiation will reach the PV modules. In regions near deserts, rain is scarce and may be accompanied by dust, which will lead to increased soiling losses.
- The existence of a high level of humidity in the atmosphere can cause fog and clouds, reducing the irradiance that reaches the PV surface. Sun radiation will contain a greater diffuse part and PV plant yield will be smaller. Condensation of humidity in the first hours of the day can affect the dust deposited on the module and increase the soiling losses.

4.4.1 Soiling (dust and dirt)

Losses due to dirt, bird droppings, dust or sand accumulation on the modules (see Figure 4.4) are included in the soiling loss factor ($L_{soiling}$). Typical values for $L_{soiling}$ are in the range of 2 to 15 per cent but can reach up to 80 per cent or more in deserts or dusty locations if PV modules are not cleaned regularly enough.

58 SOLAR PUMPING FOR WATER SUPPLY

Figure 4.4 Sand on PV modules limiting power output

Figure 4.5 Shading of PV modules limiting power output

New developments in the glass surface of PV modules are aimed at reducing the accumulation of dust. A similar effect is obtained with an increase of the tilt angle, so that there is less accumulation of sand on the PV modules, where the better $L_{soiling}$ compensates the losses due to a non-optimized tilt (see more on tilt angle in section 5.3).

> **Box 4.3 Minimizing losses due to soiling**
>
> It is important to ensure regular cleaning of PV modules (clean water and cloth, no soap needed) by the community of users or an external party (from several times per week to once a month, depending on level of dust and rainfall frequency) as otherwise excessive dust/dirt on modules might significantly reduce water output or even bring it to zero.

4.4.2 Shading

Losses due to shading are represented by the term $L_{shading}$ (Figure 4.5). Several areas of shading are present in PV installations:

- shading between rows of modules;
- shading due to nearby obstructions, such as vegetation, bushes, trees, buildings, poles, railings, antennas, signals, power lines, or overhead wiring;
- shading due to obstructions on the far horizon, such as mountains or buildings.

If enough space is available, inter-row spacing will be calculated to avoid PV modules casting shadow on others at any moment of the year. If space does not allow, specifications would detail the number of hours of sunlight in the PV plant, free of shadow in winter solstice around noon.

Nearby shading should be avoided as much as possible, since electrical production of PV modules is non-linear, so a small shadow can produce a large power loss in the system.

Obstructions on the far horizon produce shadow mainly during sunrise and sunset, when irradiance levels are lower. Losses can be elevated in deep valleys between high mountains.

The analysis of shading effects on PV modules is complex due to the non-linear behaviour of PV modules. The impact of shading on PV arrays can be evaluated by representing the obstructions on a sun position chart for the latitude of the PV installation location. The reduction in the number of peak sun hours due to shadow is calculated comparing the shadowed area with the area of the complete solar window. There are some software programs and devices that can provide a derating shading factor, simplifying the shading

> **Box 4.4 Minimizing losses due to shading**
>
> The structure supporting the PV modules must be elevated over the terrain in ground-mounted PV installations if vegetation and bushes grow quickly. Other common elements in buildings (poles, cables, antennas, etc.) should be installed on the north side of the building, so that the PV modules can be installed facing south (in the northern hemisphere), free of shadow. Including the absence of inter-row shading as a condition in the contract with the installer will help to minimize such events. Involving the community of users (or third parties) for regular clearing of the area around the PV plant (e.g. trimming of nearby trees) and ensuring no future construction overshadows the PV modules will decrease the likelihood of losses due to shading.

losses evaluation. In a first approximation $L_{shading}= 0$ per cent unless a shadow study is carried out for the particular conditions of the installation.

4.4.3 Angular and spectral reflectance

Electrical characteristics of PV modules are obtained by the manufacturer during a flash test with a light perpendicular to the PV module surface. Yet the sun moves from east to west due to the rotation of the Earth and, therefore, in fixed-tilt PV systems, the light incident angle on the PV module varies throughout the day. The difference between the reflectance of a PV module under real operating conditions and the reflectance in the flash test is quantified by the term L_{ref} or losses due to angular and spectral reflectance (also denoted as optical losses).

Advancements in the front glass are aimed at increasing light transmission using a thinner glass, improving surfaces to reduce dust accumulation, and reducing light reflection with anti-reflective glass coatings. Similar advances are being applied at the PV cell level, with new PV cell surfaces and new anti-reflective coatings intended to reduce optical losses by catching the maximum amount of solar radiation.

Typical values for L_{ref} are in the range of 2 to 6 per cent in fixed-tilt crystalline PV modules, depending on the quality of the materials used in the construction of the PV module.

The irradiance incident on the PV surface is reduced when the air is polluted. Air pollution is mainly produced by human activity and can reach significant levels in the proximity of urban and industrialized locations. Polluted air can contain fine solid particles, liquid droplets, and gases. These substances reduce the total solar radiation reaching the PV modules due to part of the direct solar radiation being diffused or reflected. These losses are highly dependent on the particular characteristics of each location and there are no typical values to quantify them.

4.4.4 Incorrect azimuth orientation and tilt angles of PV modules

Losses due to an incorrect orientation (L_{ori}) or tilt angle (L_{tilt}) of the PV plant are related to the energy yield that can be obtained with the optimum PV azimuth orientation and tilt angle that generates the maximum energy (see more on tilt and azimuth orientation angles in section 5.3)

In general, the two loss factors are calculated together with the following expression:

$$L_{ori+tilt} = \left(1 - \frac{PSH_{ori+tilt}}{PSH_{optimum}}\right)$$

The term $PSH_{optimum}$ corresponds to the irradiation for the orientation and tilt angle that maximizes the energy produced in a period and $PSH_{ori+tilt}$ is the irradiation for the orientation and tilt angle of the PV array. $L_{ori+tilt}$ is commonly calculated on a yearly basis and is highly dependent on the particular characteristics of each PV installation (see example in Annex C). Losses due to tilt angle not being the optimum value or azimuth orientation (deviation from south) are usually non-existent in most installations as it is simple to find out the best angles for a given location. As an approximation, for every 1 degree off the optimum tilt angle, the losses will be of around 0.1 per cent. Most design software from reputed manufacturers will take into consideration the losses due to non-optimal azimuth orientation and tilt angles.

> **Box 4.5 Minimizing losses due to incorrect azimuth and tilt angle**
>
> Define clearly in the tendering, bidding, and/or installation specification documents the azimuth and tilt angle that is desired. When PV modules are being installed, check in the field with the help of a compass since this is commonly the moment when most errors are made.

4.5 PV module energy losses

There are several energy loss factors related to the PV module. The cell temperature during the normal operation of the PV plant is the most important one, as explained in section 4.2. Other factors are: power tolerance, mismatching, ageing and light-induced degradation, and low irradiance efficiency.

4.5.1 Power tolerance

PV module manufacturers guarantee the module's peak power at STC within a given tolerance range. Not reaching peak power is related to the power tolerance detailed in the PV module datasheet and is influenced by the quality of the module (and the manufacturing process). The power tolerance of commercial PV modules varies greatly depending on the manufacturer: 0 per cent to +3 per cent; –3 per cent to +5 per cent; ±3 per cent; ±5 per cent; etc. The power loss due to power tolerance ($L_{tolerance}$) corresponds to the minimum tolerance detailed in the datasheet and determines the possible loss of energy yield of the PV installation in the worst case scenario.

4.5.2 Mismatching

The values of I_{MPP} and V_{MPP} (section 2.10) may vary slightly from one PV module to another, even when they are the same model of PV module. The connection of modules with a lower I_{MPP} will reduce the peak power of the PV field, since the I_{MPP} in a string of modules connected in series will be equal to the lowest I_{MPP} of any individual module in the string.

Mismatch losses ($L_{mismatching}$) appear due to differences in the parameters of the modules used in the installation. For losses due to mismatching, a value of 2 per cent is used in the design of PV installations that use crystalline modules. Manufacturers of thin-film photovoltaic modules claim that their technologies have fewer mismatch losses, with values that are close to zero. This value can vary substantially depending on the quality of the modules used in the PV installation, so it is important to connect in series those modules with very similar characteristics.

A bigger problem appears when PV modules of different manufacturers or different power ratings are mounted in the same PV plant (for example, when a contractor has run out of stock of PV modules in the middle of an installation and brings other models in, or when a PV module is broken in an existing installation and it is not possible to find the same model in the market). This situation, especially when a new PV module with lower I_{MPP} and V_{MPP} is mounted in a PV plant with modules that have larger I_{MPP} and V_{MPP} values, is to be avoided at all costs as it might affect the entire PV plant output and considerably reduce energy production and water output.

Finally, differences in cable length or cross-section among parallel strings can introduce differences in voltage drop and therefore contribute to an increase in mismatching losses.

> **Box 4.6 Minimizing mismatching losses**
>
> All PV modules must be from the same manufacturer and be of the same type and model when constructing PV fields. When some modules break down and need to be replaced by new modules the strings should be reorganized using the same model of modules in as many strings as is possible, adding the new modules to complete the incomplete strings.
>
> The new modules must be of the same PV technology (whether mono-Si or poly-Si), must have the same number of cells, and must have an I_{MPP} and V_{MPP} of the same value or greater than the modules they are replacing.
>
> In addition, cable of the same length and cross-section should be used as much as is possible to connect strings in parallel.

4.5.3 Ageing and light-induced degradation

Ageing is one of the common names given to the long-term losses caused by the PV modules being exposed to light, also known as light-induced degradation (LID). It is important to distinguish between the long-term degradation of PV modules and the initial stabilization that appears after the beginning of light exposure. Initial stabilization reduces the PV module efficiency in values in the range of –2 to –4 per cent in the first weeks of operation. The maximum losses for long-term light-induced degradation, denoted as L_{LID}, are limited by the power warranty given by the manufacturer, which can be considered as a conservative estimate for L_{LID}.

In Figure 4.6, the initial stabilization of the PV cell entails a 3 per cent decrease and the maximum annual degradation rate is equal to 0.68 per cent during the 25 year lifespan. It is important to note that the most common solar pumping design software does not consider losses due to ageing and degradation over time of PV modules.

ENERGY LOSSES IN SOLAR PHOTOVOLTAIC ENERGY PRODUCTION

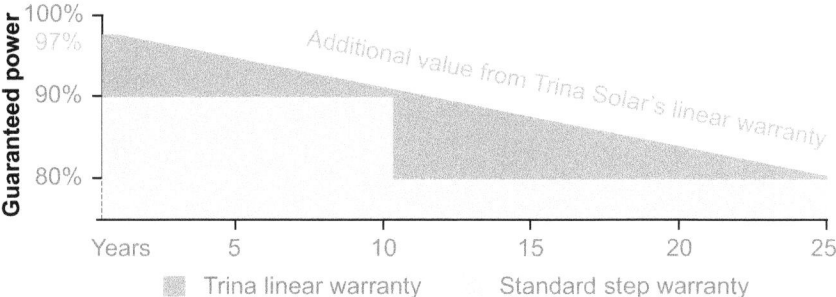

Figure 4.6 Example of warranty given by Trina Solar for Tallmax monocrystalline modules

---

Box 4.7 Minimizing ageing losses

When designing the PV installation, the power output from the PV plant at the end of the design period can be estimated taking ageing into account (e.g. looking at the manufacturer's PV warrantied power output graph). The PV field can now be oversized to ensure the required power output will be available at the end of the design period despite ageing. Care should be taken not to go beyond the maximum voltage input of the inverter when oversizing the PV field.

4.5.4 Low irradiance efficiency

The *low irradiance efficiency loss* depends on the technology and characteristics of the PV module and the irradiance level. Some manufacturers of thin-film modules state that their technologies have a better low-light performance than crystalline modules, producing more power at the nominal values at STC. Locations with fog or overcast skies have an increased portion of diffuse light reaching the PV module surface. Some crystalline module manufacturers establish in their datasheets a 4.5 per cent reduction in the module efficiency for irradiance equal to 200 W/m².

4.6 Module mounting energy losses

In addition to the above-mentioned losses, the mounting of the PV modules, whether in portrait or landscape orientation, influences energy loss.

Further to the inter-row shading losses explained in section 4.4.2, when a shadow occurs on a PV module, the shadowed cells will start dissipating the power generated by non-shadowed cells, getting hot to the point that the whole PV module may malfunction. In order to avoid the appearance of these hot spots, when PV module shadowing occurs, the bypass diodes in the junction boxes at the back of each PV module will act as a switch, disconnecting the shadowed cells from the rest of the PV module.

Usually one bypass diode protects a group of 18, 20, or 24 PV cells, which are short-circuited and do not contribute to the generation of energy. PV modules mounted in landscape configuration will typically experience less

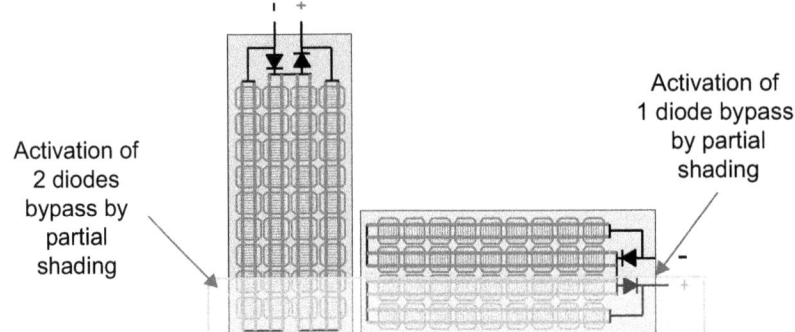

Figure 4.7 Bypass diode activation in shadowed PV modules

inter-row shading losses than modules installed in portrait configuration due to the internal connection of the bypass diodes, meaning that only one, rather than two, diodes will activate, as shown in Figure 4.7.

Thin-film modules respond differently to shading. The difference in PV cell construction means that losses due to shading will be proportional to the area shadowed, no matter whether in portrait or landscape orientation.

4.7 Power converters and the balance-of-system energy losses

Balance of system, or BoS, describes all the components of the system other than the PV modules, power converters (DC/DC control boxes or DC/AC inverters), and water pumps. The BoS includes wiring, electrical safety elements, auxiliary equipment and sensors, the supporting structure, and mounting hardware. The most significant loss factors related to power conversion and BoS elements are:

- wiring (explained in section 4.3);
- energy efficiency of the electronic power converters;
- errors in tracking the MPP;
- PV plant downtime for maintenance, breakdowns, or malfunctions;
- transformers (for grid-connected systems);
- other losses, such as those due to safety elements and auxiliary equipment.

Several power converters can be used in PV systems: grid-connected inverters, off-grid inverters, DC/DC or DC/AC converters, and so on. The main losses related to power converters are MPP tracking losses, represented by L_{MPPT}, and losses in the conversion, represented as L_{conv}.

Some power converters are needed in order to establish the PV operating point in the MPP of the PV field. The MPP tracking algorithm allows the generation of the maximum power (and energy) from the PV field, optimizing the overall performance of the system. Nowadays the losses due to MPP tracking (L_{MPPT}) are very small, less than 0.1 per cent.

During the operation of the power converters some losses appear in the semiconductors that switch high voltages and currents. The efficiency in the conversion (η_{conv}) varies with the characteristics of the pump connected to the output and the input power source. Depending on the power converter type, manufacturers can provide the efficiency profile of the converter.

The losses in the conversion, represented as L_{conv}, are calculated as:

$$L_{conv} = (1 - \eta_{conv})$$

Typical values of L_{conv} vary in the range of 1 to 5 per cent.

In addition, there are sometimes intervals in which the PV plant is not generating energy due to a failure in some part of the system: inverter disconnection, operation and maintenance (O&M) works on the PV plant, problems in the grid connection (grid availability and disruption of grid-connected systems), inverter or pump failure, and so on.

The typical availability losses ($L_{availability}$) related to grid functionality in grid-connected systems in developed countries are in the range of 1 to 3 per cent. This can get much higher for areas where the grid is unstable or where specialized technical assistance is needed but not immediately available.

> **Box 4.8 Minimizing losses due to availability**
>
> The functionality of the PV pumping scheme and expected maximum downtime and repair times should be stated in an O&M contract. O&M services could be an integral part of the installation contract with the private contractor selected for the works, and would detail the time to respond to the different failures that can appear in a PV system. See more on O&M services in section 11.2.

In some PV–grid hybrid systems it is necessary to use a transformer to adjust the levels of the AC voltages between the inverter and the electrical grid. Losses in the transformers (L_{transf}) vary depending on the quality of the materials used in the construction of the transformer, with typical values between 1 and 5 per cent, or even greater than 7 per cent in poor-quality transformers.

These days, due to the falling prices of PV modules and the need for pumped water in times of low sun irradiance, it is common to oversize the PV plant. In these cases, when the sun irradiance levels are high, the PV plant is able to produce more energy than the pump requires to work at 100 per cent of its capability. The power converter will only 'take' the energy needed by the pump, and the rest is not used or 'wasted'; that is called a curtailment of power generation. The curtailment of the power generation reduces the performance ratio of the installation, since the PR is calculated against the peak power of the PV field.

Different electrical devices are commonly used in PV installations for the protection of the installation (safety elements) or for the electrical connection: string fuses, array over-current protective devices, automatic circuit breakers, switches, and others. Losses due to these parts are difficult to calculate and are not usually considered in the estimation of the energy generated by a photovoltaic system.

4.8 Estimation of the energy yield

The estimation of the production of a PV system is carried out using the performance ratio of the installation using the following expression:

$$E_{generated} = P_{pk} \cdot PSH \cdot PR$$

The estimated value of PSH can be obtained using the PVGIS calculation tools available at the European Commission's Photovoltaic Geographical Information System[1] or other websites. The PR of an installation takes into account all the power losses that can be determined for the PV system, such as the factors described in the previous sections, calculated as:

$$PR = \begin{bmatrix} (1-L_{tem}) \times (1-L_{wires}) \times (1-L_{soiling}) \times (1-L_{shading}) \times (1-L_{ref}) \times (1-L_{ori}) \\ \times (1-L_{tilt}) \times (1-L_{tolerance}) \times (1-L_{mismatching}) \times (1-L_{LID}) \times (1-L_{MPPT}) \\ \times (1-L_{conv}) \times (1-L_{availability}) \times (1-L_{trans}) \end{bmatrix}$$

A plant with a high PR converts solar radiation more efficiently into electrical energy. PR is usually expressed as a percentage and presents seasonal variations according to environmental conditions through the year. In addition, a degradation factor of –0.5 per cent/year (approx.) must be taken into account over the lifetime of the installation. The PR of PV systems depends strongly on the type of installation and the components used in it.

The PR in a well-designed off-grid PV system is in the range of 0.5 to 0.75 a value closest to 1 is preferable (excluding energy losses in the water pump).

Equating the $E_{generated}$ with the $E_{required}$ by the pump to operate for a defined number of hours (P_1 of pump motor x no. of pump working hours, see example in Annex B), it will be possible to get the required P_{pk} of the solar PV scheme.

Conversely, the PR can be calculated in real PV installations if the other terms of the expression are known:

$$PR = \frac{E_{generated}}{P_{pk} \cdot PSH}$$

Monitoring systems included in PV installations allow the measurement and recording of the energy generated and the irradiance. PSH is calculated from the irradiance values. The PR allows the comparison of PV plants over a given time independent of PV plant peak power or solar resource. A high PR is representative of a well-built and well-functioning installation.

Note

1. https://re.jrc.ec.europa.eu/pvg_tools/en/tools.html

CHAPTER 5
Design of a solar-powered water scheme

The high variability of factors influencing water output in solar-powered schemes makes it difficult to optimize sizing when solar design software is not used. Solar software developed by different manufacturers is increasingly popular and available. Yet the clarity of the data required beforehand and good design criteria are key to coming up with an efficient design. These are explained in detail in this chapter together with the different secondary energy options whenever a design leads to hybrid configurations.

Keywords: solar design software, module orientation, solar pumping layout, parallel and series configuration, solar design month, solar tracking

5.1 Solar pump design

As explained in previous chapters, there are numerous factors that influence the production of electricity through solar modules, and therefore the water output in a solar pumping scheme. Most of those factors are variable with the time of day and season, making it complicated and imprecise to proceed to design a solar pumping scheme using manual calculations.

Numerous software packages that will facilitate this process are available (e.g. Grundfos, Lorentz, Wellpumps), computing all factors for the chosen components and geographical locations, and proposing designs including solar module layout and power, cable sizes, inverter or control box models, pumps, and balance-of-system components. Software-based solutions will also match performance and electrical characteristics of the components, to ensure expected electrical and water outputs as much as possible.

Whether a field technician decides to come up with the design or commissions a private contractor, the authors of this book believe the design should always be carried out through a reputable software design package.

However, no matter how good and user-friendly the software is, it is always of paramount importance to understand what data are needed for the design, and how different choices will affect the system performance.

Most of the steps involved and data needed to design a solar water pumping scheme are common with those needed to design a generator-powered water scheme. A summary of the data required is given in Table 5.1.

http://dx.doi.org/10.3362/9781780447810.005

Table 5.1 Data needed to design a solar-powered water scheme

Data	Comments
GPS location (degrees) of water source	Precision of seconds or more than one decimal is not needed
Maximum daily water requirements (m^3/day)	Bear in mind that demand may grow over the design period of the water scheme and the seasonality of water needs
Safe yield of water source (m^3/hour)	To be determined after pumping test (or pumping yield to match if retrofitting an existing system)
Expected maximum drawdown (m)	Determined from nearby water sources (seasonal drawdown) and/or hydrogeological reports
Static and dynamic water levels (m)	Determined from pumping test
Inside casing borehole diameter (inches)	From borehole drilling report
Depth of pump installation (m)	Needed to calculate losses in pump to inverter cable
Distance from water source to solar modules (m)	Typically, this should be no more than a few metres from the water source
Distance from water source to tank (m)	From topographical assessment
Elevation difference between water source outlet and tank inlet (m)	From topographical assessment
Space available for solar modules (m^2)	Field visit to the area
Pipe type from pump to tank (metal, plastic)	Also pipe size if already existing
A list of equipment characteristics (in addition to the above) in case of solarizing an existing water scheme	Generator size, water pumping yield, pump type and model, motor power rating, cable wire size, pump depth installation, pipe materials and size, size of tank

5.2 Important design concepts and considerations

In designing a solar pumping system, there are several important concepts to take into consideration.

- *Design is based on water pumped per day as opposed to water pumped per hour.* Unlike traditional pumping, which is possible 24 hours a day, with solar pumping there is a limited number of hours in a day available for pumping, a period typically called the solar day. With this concept, the total water requirement for the whole day is factored in and a pumping system that will meet that requirement during the solar day is selected.
- *Batteries to store energy are to be avoided.* Batteries introduce inefficiencies (power losses), are expensive (increased system costs), heavy (difficult delivery), have a relatively short life (replacement costs), and introduce complexity, such as how to dispose of them at the end of their life

cycle. Instead of storing energy in batteries, it is recommended to store water in elevated tanks (which is another form of storing energy), making water available when needed and keeping the system simple and efficient.
- *The tank storage capacity must be factored in.* The size of the storage tank should be determined by the daily demand and should be large enough to store as much water as possible during daylight hours. In areas that have significant seasonal variations in solar radiation (due to weather) the tank would be larger than in areas where there is less variation (see more in section 5.3.8).
- *Solar tracking is discouraged.* Due to the cost of tracking systems, their operation and maintenance requirements, and plummeting PV module prices, solar tracking is not considered especially useful. It is more common and easier to oversize a fixed PV generator to cater for losses due to orientation of the PV array relative to the sun than to incur the cost of more complex PV tracking (see more on tracking in sections 5.3.5 and 7.3).

5.3 Steps to design a solar-powered water scheme

In order to obtain the required data, a number of steps need to be taken to ascertain whether a stand-alone solar system can meet the required water demand from the selected water source or a hybrid (solar + one or more additional power sources) is needed. These steps are:

1. Water-demand assessment and design period.
2. Water-source assessment, borehole construction, and pumping test.
3. Design month and flowrate.
4. Calculation of total dynamic head and pump selection.
5. Solar PV array sizing, layout, inverter and module selection, and orientation of solar panels.
6. Location of main components.
7. Size and location of water storage tank.
8. System layout: stand-alone solar vs hybrid systems.
9. Balance-of-system minimum requirements.

5.3.1 Water-demand assessment and design period

As for any other water-scheme design process (not specifically solar), the water demand of the target population must be estimated. In order to do so, a project design period will need to be established. This is the period of time in which the solar system must provide the amount of water set as the target. Water demand typically grows over time as population and other needs (e.g. irrigation, livestock) increase.

Ideally the system will be designed to meet water demand in any given month of the design period. The longer the selected design period is, the higher the water requirement to be met typically will be and therefore the more oversized (compared to the current situation) and the more expensive the whole water-scheme system will be. In practical terms, the length of the design period will mostly depend on the budget available and will be a question of investment priority.

The typical lifetime of quality components can serve as an additional reference and these are given in Table 5.2:

Table 5.2 Expected lifetime of solar components

Component	Expected lifetime (years)
Borehole	20–25
Pump	5–10
Inverter/control box	5–7
Solar module	25
Civil infrastructure (water tower, module supports)	20–25
Water tank – cement	15–20
Water tank – plastic	5–10

Since it is difficult to know how the water demand, population, and other factors will grow and influence the system, it is often argued that it is impractical to have very long design periods (over 20 years), which might be especially true for humanitarian operations where donors have short funding periods, funding is often restricted, and where contexts are volatile and might dramatically change over the years. It also has to be taken into account that a design period of 10 or 15 years will not mean that the water system will stop working after that time.

In light of the expected component lifetimes, a resizing of the system may be possible every 5 to 10 years, when the pump and inverter will typically need to be changed and could be replaced by larger models if necessary (since it will normally be possible to add more modules and increase water storage). Therefore, design periods longer than 5–10 years may be difficult to justify in humanitarian contexts.

5.3.2 Water-source assessment, borehole construction, and pumping test

While both surface and submersible pumps can be solarized, this chapter will largely consider groundwater since it is more widely used due to wider availability and better quality of water. Surface sources (rivers, ponds, lakes) can also be exploited through solar pumping technology and the application principles explained in this chapter remain the same.

Before installing any water pumping scheme, an understanding of the water resource is needed; for groundwater aquifers, it will be the work of hydrogeologists to rigorously assess groundwater availability, storage and recharge of the aquifer as well as water quality. This will be particularly relevant in areas where an increasing number of boreholes are drilled, in order to ensure a sustainable use of the resource.

Once borehole siting and drilling (or rehabilitation of an existing one) is done (for more detail, see ICRC, 2010), a test pumping is carried out to determine the unique characteristics of the water source, including the tested yield and the dynamic water level that are critical for properly designing a pump to match the borehole. Test pumping includes a 4-hour step-down test followed by a constant rate test for 24–72 hours. The point at which the rate of pumping out of the borehole remains constant even with continued pumping is recorded as the tested yield. This is also the point at which the water level in the borehole does not change and is recorded as the dynamic water level (also known as pumping water level).

The safe yield is the maximum amount of water that can be safely extracted from the borehole without compromising its integrity. It is a percentage of the tested yield of the borehole and is a value between 60 and 70 per cent of the tested yield.

Many aspects affect the safe yield of a borehole, namely:

- borehole siting and groundwater exploration (the 'natural' geological/hydrogeological conditions);
- borehole construction and development (the 'human-made' conditions acting on the hydraulic behaviour of the well);
- drilling and pump test (see more at MOAIWD, 2012) supervision and documentation (which act on the resulting 'safe yield' interpretation).

In the absence of clear, detailed documentation of borehole characteristics and pumping test results, the design and equipping of a pumping scheme should not be done as there is a high chance of over or underestimating yields, leading to malfunction, over-pumping, the borehole drying up, and a number of other problems. This is a critical mistake that is often encountered in field humanitarian projects.

> Assessment of the water point, including pumping test results, must be carried out before any design takes place.

In these cases, measures should be taken to find out the main characteristics of the borehole, including carrying out a pumping test for which a contractor might be hired.

In the case of retrofitting an existing water scheme to solar, assessment of the borehole and its characteristics and safe yield (following the above-mentioned studies and tests, especially pumping tests in case of doubt) must be done before proceeding to any solar design or installation.

Finally, it should be kept in mind the date at which studies and tests are carried out and how seasonality may affect the water-source level (e.g. through monitoring water levels in other boreholes situated in the same aquifer). For example, in some areas of the Middle East, drawdown during a dry season might be over 100 metres; not taking this into consideration at the time of designing a pumping scheme might lead to undersized equipment being installed and other problems, resulting in the required water not being provided during part of the year.

5.3.3 Design month and flowrate

In solar pumping schemes, and since solar irradiation will vary over the course of the year, it is necessary to define a design month (the month taken as a reference to size the system).

Typically, the design month when the water is for human consumption will be the one with the lowest ratio between solar irradiation at a predetermined tilt angle (see more about tilt angles in section 5.3.5) and water needs, also called the worst month. The principle behind taking the worst month as a reference point in designing the solar pumping scheme is that if it is possible to provide the water required in that month, then it is ensured that the solar system will be able to provide the required water during any other month of the year.

In order to identify the worst month for a particular location and water scheme, two situations are possible: 1) that daily water required is constant during the year; or 2) that the water required changes from month to month.

When the water required is constant throughout the year, the month with the lowest PSH/water required ratio will always be that with the lowest solar irradiation (lowest number of peak sun hours). In the example given in Table 5.3, it is the month of July.

In the example of variable requirements given in Table 5.4, more water is needed during the dry season (water use per person is typically higher, there

Table 5.3 Example of worst-month calculation where constant water required: Yumbe, Uganda (3N, 31E)

	Jan	Feb	Mar	Apr	May	Jun	Jul	Aug	Sep	Oct	Nov	Dec
Peak sun hours (PSH)	6.8	6.8	6.0	5.4	5.0	4.5	4.2	4.6	5.4	5.5	5.9	6.5
Water required (m^3/d)	120	120	120	120	120	120	120	120	120	120	120	120
PSH/water required	.057	.057	.050	.045	.042	.038	.035	.038	.045	.046	.049	.054

Table 5.4 Example of worst-month calculation where variable water required: Yumbe, Uganda (3N, 31E)

	Jan	Feb	Mar	Apr	May	Jun	Jul	Aug	Sep	Oct	Nov	Dec
Peak sun hours	6.8	6.8	6.0	5.4	5.0	4.5	4.2	4.6	5.4	5.5	5.9	6.5
Water required (m^3/d)	154	165	140	120	110	105	98	110	120	120	130	155
PSH/water required	.044	.041	.043	.045	.045	.043	.043	.042	.045	.046	.045	.042

might be irrigation schemes connected). Therefore, the worst month (the lowest ratio of solar irradiation to water required) is February.

Most of the design software available will allow the introduction of design month or give an option to proceed with the design considering the worst month as the design month.

Once the design month is known, the required pumping rate can be roughly estimated by dividing the water required by the PSH.

For example, in the case presented in Table 5.3 this is 120/4.2 = 28.5 m³/h. If the safe yield of our well is higher than 28.5 m³/h, then it will be possible to provide all the water required during the whole year with a stand-alone solar pumping scheme, and this could be the selected layout.

Other considerations beyond the technical may factor in selecting stand-alone or hybrid configuration, as discussed in section 5.3.8.

Even if safe yield is below 28.5 m³/hr, for most parts within the sun belt (the area between latitudes 40°N to 40°S) it will be possible to pump water between 6 and 9 hours per day by oversizing the solar PV array (design software can provide this precise information). This would mean that for our example of constant requirements, if the safe yield is between 13 m³/hr (120/9) and 20 m³/hr (120/6), it is still possible to provide the water required for the worst month (depending on solar irradiation at the geographical location) and hence for all the other months and, therefore, the selected layout could still be a stand-alone solar system. For safe yields under 13 m³/h, another power source (grid, generator) would be needed (or a second borehole drilled) in order to meet the required water needs.

5.3.4 Calculation of total dynamic head and pump selection

Calculation of total dynamic head (TDH) and subsequent pump selection is a step that can be done independent of the power source to be selected and is essentially no different than if the pump is powered with a diesel generator.

Once the desired hourly pumping flow has been defined and the TDH of the system calculated, H-Q curves for different pumps will be reviewed in order to find a suitable one (numerous pump manufacturers' catalogues are found online with H-Q graphs for all their pumps). Care should be taken to select a pump with a nominal flow for the given TDH that is close to the desired flow. A step-by-step example of this process can be found in Annex A.

Whenever field conditions and requirements to be met allow, the choice of a DC pump over an AC pump is recommended for solar stand-alone pumping schemes. While DC pumps will generally be a bit more costly than their equivalent AC ones, they have a longer lifespan, they are more efficient (so fewer PV modules are needed to power them), and they don't need DC/AC electricity conversion, making their control box a cheaper accessory than DC/AC inverters.

When selecting a pump, ensure that the nominal flow of the pump is close to the required design flow. If a greatly oversized pump is selected, while it will still supply the water required, it will also need a greater PV generator and inverter increasing the total cost of the system. Moreover, pumps working at a small fraction of their capacity will have their lifespan shortened and will also compromise the life of the borehole, therefore increasing the capital and replacement costs of the system.

Most solar pumping design software will size the pump and offer different solutions even when TDH has not been computed, by providing other field data (e.g. dynamic water level, tank elevation, and pipe material to be used).

5.3.5 Solar PV array sizing, layout, inverter and module selection, and orientation of solar panels

Solar PV array sizing. Once the pump has been selected it will be possible to estimate how much energy needs to be generated to power it.

Section 4.1 showed that for any type of PV pumping system the energy generated by the PV generator can be estimated by the following expression:

$$E_{generated} = P_{pk} \times PSH \times PR$$

Section 4.8 showed that the energy required by the pump ($E_{pump} = P_1 \times$ hours of operation of the pump) can be equated to $E_{generated}$ to get the required P_{pk} of the solar PV scheme.

$$E_{pump} = E_{generated} = P_{pk} \times PSH \times PR = P_1 \times hours\ of\ operation\ of\ the\ pump$$

Once the power of the pump (and hence the energy requirements), the PSH in the location, and an estimated PR are known, it will be possible to calculate the P_{pk} (peak power of the solar array) and therefore the minimum number of PV modules needed (considering the models available in the local market). An example is given in Annex B.

The higher the power rating of modules, the fewer will be needed. Silicon crystalline modules of up to 500 Wp can be found on the market and power ratings are constantly growing.

Solar PV array layout and inverter selection. Once the number of modules is known it will be important to define the configuration; (deciding how many are to be mounted in series and how many in parallel) so that not only the power, but also the voltage, and current provided by the solar array are right for the inverter and pump.

When PV modules of same model and ratings are mounted in series, voltage adds up and current remains the same, and vice versa when mounted in parallel.

PV modules are connected *in series* (connecting the positive terminal of the first to the negative terminal of the second module and so on, as shown in Figure 5.1) to generate greater voltage. A series connection increases voltage but the current remains unchanged. In other words, the output voltage, V_{mp}

of a series connection is the sum of each individual module voltage in that connection, while the output current is that of a single module.

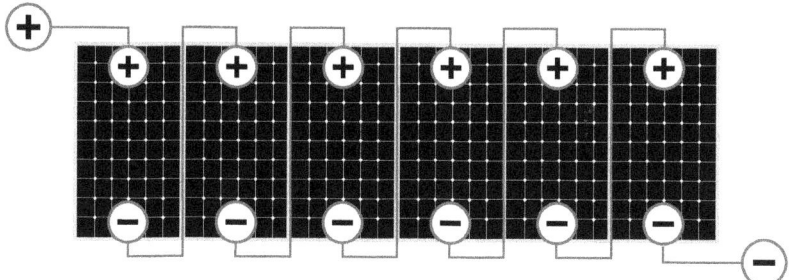

Figure 5.1 Series connection

Referring to the module characteristics given in Table 3.2 where

$$P_{max} = 250\,W,\ V_{mp} = 30.4\,V,\ \text{and}\ I_{mp} = 8.23\,A$$

Power = 250 × 6 modules = 1,500 W, Voltage = 30.4 × 6 modules = 182.4 V, and Current = 8.23 A

PV modules are connected *in parallel* to generate greater current. A parallel connection means connecting all the positive terminals together and all the negative terminals together, as shown in Figure 5.2. A parallel connection increases current, but the voltage remains unchanged. In other words, the output current of a parallel connection is the sum of each individual module current in that connection, while voltage is that of a single module.

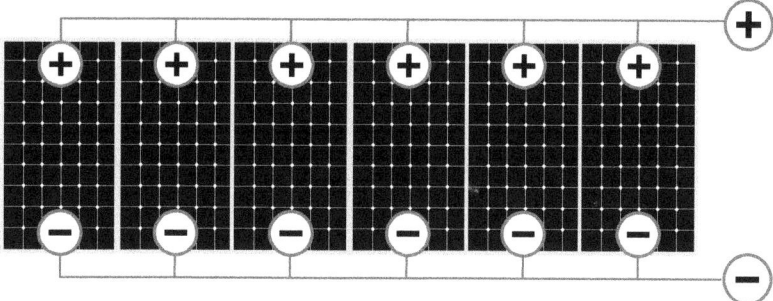

Figure 5.2 Parallel connection

Power = 250 × 6 modules = 1,500 W, Voltage = 30.4 V, Current = 8.23 × 6 modules = 49.38 A

Parallel and series connections can be combined to increase both the voltage and the current according to what is required by the pumping system.

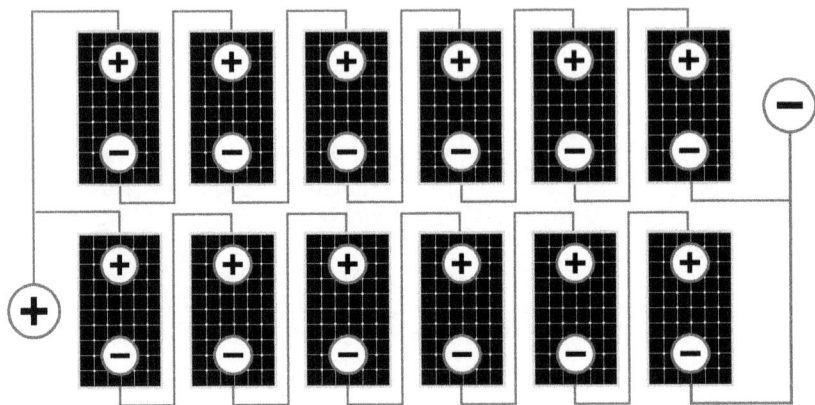

Figure 5.3 Combination of series and parallel connections

Power = 250 × 12 modules = 3,000 W, Voltage = 30.4 × 6 modules
= 182.4 V, Current = 8.23 × 2 strings = 16.46 A

Note that the power output of the connection is the sum of the power of each individual module in the connection regardless of whether the connection is in series or in parallel.

These calculations are only valid if all the modules have the same voltage and the same current. If the panels in series do not have the same current, the array will be limited to the lowest current value of the individual modules. Likewise, if the modules in parallel do not have the same voltage, the array will be limited by the lowest voltage value of the individual panels.

Parallel connections increase the current in the system, hence requiring thicker cables to carry the current (refer to sections 2.9 and 3.4.4).

Deciding how many modules are mounted in series and in parallel will be of paramount importance since it is not only the total power of the PV array

---

Example 5.1

Eighteen PV modules with characteristics as given in Table 3.2 have been interconnected to operate a water pumping system. The array consists of six modules in series and three strings in parallel (6s × 3p). The PV array will have the following parameters:

$I_{mp} = 8.23 \times 3 = 24.69 A$, $V_{mp} = 30.4 \times 6 = 182.4 V$, $P_{peak} = 250 \times 18 = 4,500 Wp$

maximum array current $(I_{sc}) = 8.81 \times 3 = 26.43 A$

maximum array voltage $(V_{oc}) = 37.6 \times 6 = 225.6 V$

These values correspond to the electrical characteristics under standard test conditions – AM 1.5, irradiance = 1.0 kW/m² – and the operating temperature (T) of each module is 25 °C. In the real world, array output is expected to drop, as discussed in section 2.7 and Chapter 4.

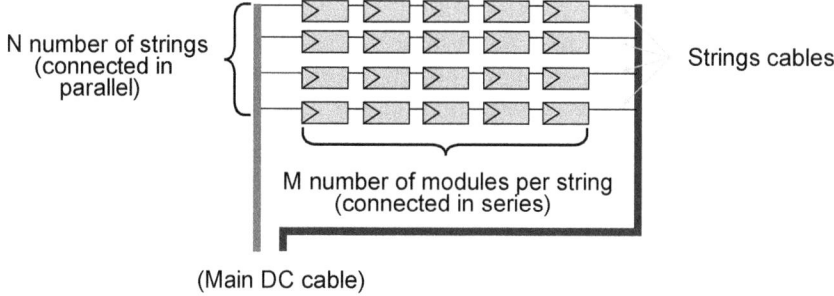

Figure 5.4 Module array layout

that matters, but also the voltage and current that the pump will get from the PV modules. A poor layout may mean that the pump will give less water (or none at all in extreme cases).

The controller/inverter to be used influences the series/parallel configuration. The selection of an appropriate controller/inverter is based on the pump selected and the PV generator size. The controller/inverter should be able to handle the incoming DC power from the PV generator and condition it to that which is required by the pump.

A manually calculated example can be found in Annex B. Fortunately, the recommended layout of the PV generator (number of modules to be connected in series and parallel) for a predefined PV module and the appropriate inverter size are provided by most reputable solar design softwares on the market (including Lorentz and Grundfos), ensuring optimum configuration and matching of PV module electrical characteristics for the given pump and water needs.

Solar PV module selection. There are several kinds of solar modules that could be incorporated in a solar water pumping design (see more in section 3.3.3). As long as the brand of the module meets the required quality certifications (see section 10.6), the module can be considered fit for purpose. Having PV modules with quality manufacturing certifications should always be the first aspect to consider and those without certifications should not progress through the procurement process.

In the great majority of cases, mono or polycrystalline modules will be used for the solar generator as they account for around 95 per cent of the market and are available in nearly all national markets around the world. Choosing one or the other will be a question of space for installation, price, and stock availability. In cases where ambient temperatures are high, modules with a lower temperature coefficient will have fewer losses and are preferred (see more in sections 4.2 and 7.8.1), so monocrystalline will behave slightly better in hot climates.

For average temperatures maintained over 40°C, amorphous silicon modules can also be considered; these will take up to three times more space because their efficiency is significantly lower, but the losses due to heat are also considerably lower than mono or polycrystalline modules.

Some design software (e.g. Wellpumps, Grundfos, Lorentz) allows users to manually introduce the characteristics of any module and would generate a design using that particular module.

Other considerations at the design stage are the mounting of the solar modules (ground, on poles, on rooftops). The advantages and disadvantages, and consideration of orientation and inclination of modules are given in the next subsection and in section 6.2.5.

If for any reason, the ideal orientation and inclination of modules cannot be ensured (e.g. modules are mounted on the roof of an existing building and therefore mounting angles are predetermined by the roof), oversizing the solar array may be necessary. Some design softwares (e.g. Lorentz) allow for inserting different tilt and orientation angles and will propose different solar generators depending on those.

Modules could be given another orientation if the designer decides that is a better option for a particular site. For example, if more water is needed during the mornings the designer might opt to orient the modules more towards the east, knowing that in the afternoon water output will be reduced. The more modules oriented to the east, the more water will be pumped during the morning.

Orientation of solar panels. Available PSH will vary (be higher or lower) depending on how the PV module is oriented with respect to the sun. The sun's path in the sky changes over the year; during summer, when days are longer, the sun's path is higher in the sky, while in winter, with shorter days, the sun's path is lower in the sky, as illustrated in Figure 5.5.

The sun's position in the sky with respect to the earth's surface is defined by three angles: azimuth angle, zenith angle, and elevation angle.

The azimuth defines the daily movement of the sun from east to west and is the angle on the horizontal plane between the projection of the beam radiation and the north–south direction line. The elevation angle defines the north–south trajectory of the sun through the seasons and is measured in degrees from the horizon of the projection of the radiation beam to the position of the sun. The zenith angle is the angle of the sun relative to a line perpendicular to the earth's surface. These angles are represented in Figure 5.5.

Maximum energy of the sun will be obtained on a PV module oriented at right angles with the sun (see Figure 5.6). This can only be achieved if the module's mounting structure follows the movement of the sun, that is, it tracks the sun. Sun trackers follow the daily and seasonal sun path to maximize the energy yield. Trackers can follow a single axis (seasonal changes) or two/dual axes (daily and seasonal). Trackers are rarely used for solar water pumping, as their absence is easily compensated for by oversizing

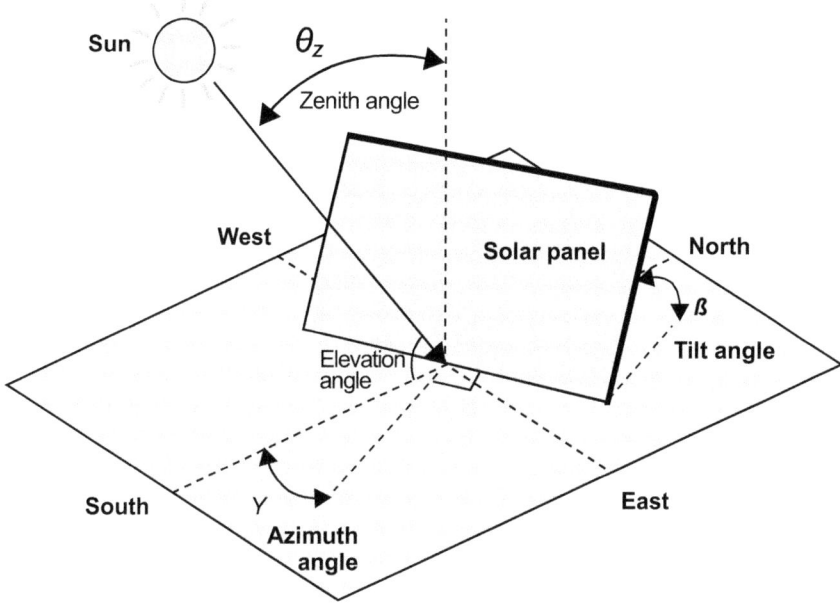

Figure 5.5 Sun's position in the sky with respect to the earth's surface

the solar PV array to account for losses due to non-optimal tilt (see Annex C for an example). Their use is discouraged as they also introduce additional complexity, involve extra costs and need specilized regular maintanence (mechanical tracking).

When trackers are not used the PV generator is mounted on a fixed structure at a fixed azimuth and tilt angle aligned so that the PV generator is at right angles to the sun. The PV array tilt angle represents the angle the array surface makes with the horizontal plane/earth. The PV array azimuth angle represents the angle between true north and the direction the array faces. With fixed structures the maximum energy – as a yearly average – is generated when the PV array tilt is equal to the latitude of the location

Application note

As a general rule for water pumping, the configuration that will yield the highest amount of energy and therefore the highest amount of water – as a yearly average – is the one where PV arrays are tilted to an angle equal to the latitude of the location where it is being installed (with a tolerance of +/–5°, which won't have a significant impact). It is also common practice to allow for a minimum tilt angle of 15° in locations of lower latitudes to allow for self-cleaning of the modules when it rains.

With regard to azimuth angle, the modules should be inclined to face towards the equator; in the northern hemisphere, modules should be inclined to face south, and in the southern hemisphere, modules should incline to face north.

80 SOLAR PUMPING FOR WATER SUPPLY

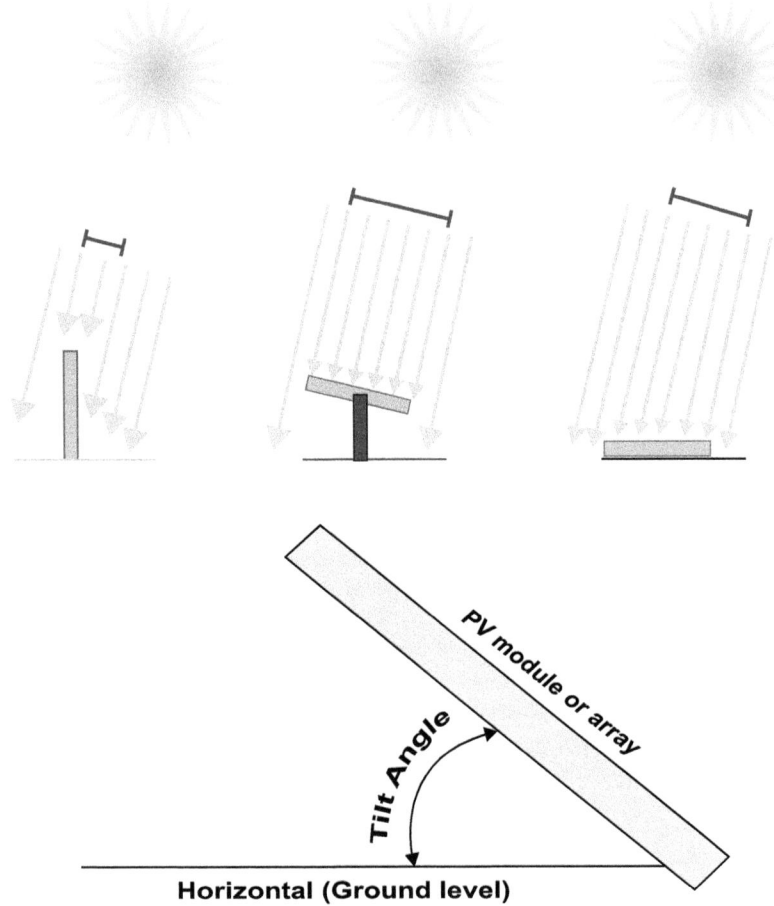

Figure 5.6 Effect of tilt on solar energy capture

where the PV array is located, and the azimuth angle is oriented so that the PV array faces the equator.

Since the height of the sun in the sky changes during the year (lower in winter, higher in summer) the tilt angle can be chosen so that maximum energy production is guaranteed in the critical season (for example, a flatter tilt to maximize water pumping in the summer when needs might be higher).

5.3.6 Location of main components

While the location of the pump will follow the same principles as in generator-powered systems (see section 6.2.2) and BoS component locations will be given by their function (e.g. dry-run sensors will be located in the borehole

Figure 5.7 PV modules in Sheikhan, Iraq, facing south at 36° tilt

just above the pump), the location of other system parts are often up to the designer to choose.

Location of solar PV modules. PV modules should be located as close as possible to the water pump (typically a few metres away from the borehole in a submersible system) in order to minimize cable losses. In all cases, they should be easily accessible for cleaning (see Figure 5.7). Modules should generally be situated in a secure and safe location as they are typically the component that is most prone to theft or vandalism.

Location of the inverter/control box. For the same reason as above, the controller should be placed as close as possible to the modules and the pump, though not in the sun as higher temperatures will lower their efficiency. Typically, it will be placed beneath modules or in a building a few metres away in order to keep it in the shade.

Location of the water tank. Locating the water tank close to the water point will minimize the number of PV modules needed to power the pump because the total dynamic head in the pumping side of the water scheme will be lower. However, this might imply the construction of a higher water tank tower if taps are located far from the water point, and vice versa if the water tank tower is far from the water point but close to the taps. A design for the same water point with the water tank located near or far from the water point could easily be run with any solar design software to appreciate the difference. Cost implications and ease of operation and maintenance of the system should be taken into account by the designer to decide the location.

Figure 5.8 Inverter box located under modules in Somali Region, Ethiopia

5.3.7 Size of water storage tank

The capacity of water storage for a solar pumping system depends on daily water use, demand, and daily and seasonal variations of solar radiation. The storage capacity should be designed to ensure water system adequacy, reliability, and compatibility with existing and future demand.

Traditionally, the size of a water tank is calculated according to the largest difference between accumulated hourly supply and demand for the month when this difference is the largest. For generator or grid-connected systems, as water can be pumped at any time (supposing grid and/or fuel supply is reliable), the volume of the water tank tends to be minimized, accounting typically for 0.5 to 1 times the daily water supply.

In solar-powered systems, however, since the time of pumping is limited to certain hours during the day, larger water tanks are normally needed. The largest amount of water pumped will normally occur during the central hours of a solar day, when radiation is highest, but during that time, water demand is commonly lower.

A larger water tank will store all the water pumped during the central part of the day so it is available when water demand is higher than supply, usually in the early morning and late afternoon. An even larger tank would help to cater for deficiency in the normal expected values for solar radiation (e.g. with longer than expected cloudy periods), adding extra security to the system.

Field survey data indicates that many SPWS storage tanks are too small, leading to water tank overflows, wasting water in the daytime and leaving shortages in the evening.

DESIGN OF A SOLAR-POWERED WATER SCHEME

Table 5.5 Quick guidance for water tank capacity

Water scheme type	Water reservoir capacity
Generator/grid powered	0.5–1 × daily water requirement
Hybrid (solar + generator/grid)	0.5–3 × daily water requirement
Stand-alone solar	1–3 × daily water requirement

A tank that is too small may also force the operator to stop the solar pumping system during the afternoon, when the highest amounts of water could be pumped.

In a hybrid (solar plus generator) system, if more water pumped from the solar source can be stored, the generator will be needed for less pumping. In practical terms, investing in larger tanks for hybrid systems is considered cost-efficient as capital costs are quickly recovered by shorter generator operating times and hence lower fuel consumption.

There is no clear formula to arrive at the water tank volume for a solar water system and an engineer will have to decide based on a compromise between present and future water demand and supply, physical space, and budget available as well as practical aspects concerning the operation of the water scheme (e.g. absence of other water points in the area, criticality of the system, and water collection patterns).

As an indication and in the absence of previous experience or data from similar projects, the values given in Table 5.5 could be considered.

From field experience, a tank more than three times the volume of the daily water requirement is normally considered impractical, mainly for reasons of budget, physical space, and/or the adverse effects of keeping chlorinated water stored for longer than three days.

Sometimes the number of consecutive no-sun days at a particular location, data that can be retrieved from solar databases such as NASA's, is taken as a reference to size water tanks. The largest number of consecutive no-sun days during the year is then taken as the water tank volume needed to cater for that worst case scenario. Since most of the time, even in countries and locations with high solar radiation, the number of consecutive no-sun days is higher than three (and in many cases between 4 and 6), the authors consider this impractical and difficult to justify, especially in the context of humanitarian operations.

5.3.8 System layout: stand-alone solar vs hybrid systems

As explained in section 5.3.3, whenever the safe yield of the water source is higher than the required yield to meet the daily water needs in the least favourable (worst) month of the year by simply pumping during the solar day, then the system layout could be a stand-alone solar system. This layout is the most cost-effective, easy to operate and maintain.

If, on the other hand, the selected water pumping rate needed to meet the water needs by just pumping during the solar day is higher than the safe yield of the water source, a stand-alone solar system won't be able to meet the demand. In this case, there are three actions that can be taken:

1. Reassess the water needs and the design period of the scheme; consider the possibility of including water-saving measures (e.g. drip irrigation) and evaluate the criticality of not meeting the required water quantity together with the community of users.
2. Have a hybrid system by including a second energy source in the water scheme (e.g. diesel generator, grid if existent) which can complement pumping beyond the solar day or for the cloudier months or days when solar pumping cannot meet needs (see Annex A for generator selection process). This can also be the case if extra security is needed to ensure daily provision water a second energy source is introduced as a back-up.
3. Evaluate the use of a second water source (e.g. drilling another borehole) that could complement the water supply. This option is especially advisable if a second energy source cannot be introduced or it is too complex or costly (e.g. in some places introducing a diesel generator may not be feasible if the community cannot pay for diesel, or diesel has to be transported from long distances).

The different pumping technologies that could be used with solar for a hybrid system are briefly discussed below.

Solar and diesel generator. This is a common layout in humanitarian operations (see Figure 5.9), especially for high-yielding boreholes or critical ones. The installation of a diesel generator will not only provide the opportunity to increase pumping whenever needed, but it also makes the system more robust if the solar parts or generator develop technical problems.

In locations with highly marked seasonality, the solar PV array needed to meet water demand in the cloudiest months of the year will have to be greatly oversized. For those locations, sometimes the installation of a generator is chosen instead. However, the way to optimize cost-efficiency in the design of a hybrid solar–generator scheme is to reduce fuel consumption by minimizing hours of generator run time. This will reduce fuel costs as well as maintenance costs for the generator.

Furthermore, since the cost of solar modules has decreased considerably over the last few years, it is technically and economically justified to design a system for 100 per cent operation from the photovoltaic array and use the generator only during certain weeks of the year where cloudy periods and bad weather are more prominent.

The final recommendation should be made based on system optimization and detailed economic analysis together with users' preferences.

Solar and grid. Grid-connected solar systems are an optimal solution for hybrid configuration. Moreover, in some locations this configuration may offer the

Figure 5.9 Hybrid solar-generator scheme at Adjumani refugee settlement in Uganda
Source: Oxfam, 2016

possibility to sell back to the grid the excess solar power generated, using the solar array as a cash crop that may result in higher financial sustainability of the whole system.

In many places where relief projects are carried out however, the grid is either non-existent or unstable, leading to strong voltage variations that could damage equipment or power cuts that could suddenly leave users with no access to water. Unless water can be supplied to communities in alternative ways whenever the grid is not functional or stable, the hybrid solar–grid option should be carefully considered.

Solar and wind power. Apart from the high unpredictability of wind power, wind technology requires specialized knowledge and spare parts with regular maintenance which are hard to find and sustain in many places where relief operations take place. Although this technology came to be highly popular at one time, these days it is less and less utilized for water pumping purposes, with its use in relief operations found to be marginal.

Solar and handpump. When borehole size allows, small solar pumps can be installed together with handpumps (both in the same borehole, one below the other). This solution saves users the burden of hand pumping and still offers the possibility to get water if the solar technology develops problems or water is needed beyond the solar day.

Care should be taken to know the safe yield of the borehole as solar pumps will typically end up extracting more water than handpumps.

5.3.9 Balance-of-system minimum requirements

The balance of system is explained in section 3.4, which details the components any solar pumping system should incorporate. Other BoS components are optional and will depend on the degree of security, monitoring, and control over the system that the designer wants the user to have.

CHAPTER 6

Electrical and mechanical installation of solar-powered water systems

The use of quality components is essential to get the expected water outputs. This chapter identifies the different electrical and mechanical parts of a solar pumping scheme and their correct installation sequence. In addition, it describes the basic electrical protection measures required to ensure safety of the water scheme and operators. Finally, a set of quality control measures together with the different options for solar module mounting are discussed.

Keywords: solar installation checklist, module mounting, solar pole mounting, Lightning Protection, Surge Protection, cable splicing, dry run protection

6.1 Pumping system installation

Quality equipment will be supplied in original manufacturer packaging and the cartons well labelled with brand names and models. The packaging should be checked for damage and to ensure it contains all the contents as ordered.

As a standard, the equipment must have a nameplate which can be counter-checked against what was ordered (see Figure 6.1). The nameplates will contain all the important data such as equipment model and power rating. A packing list comes in handy for verifying delivery of all the required equipment.

As discussed in Chapter 10, it is important to vet equipment suppliers. Nameplates and barcodes can be forged and if suppliers are not vetted to ensure they represent and supply genuine manufacturer products there is a risk of buying counterfeit equipment.

It is mandatory that before commencing the installation all supplied components be inspected to ensure they do not have any obvious damage (damage can occur in transit) and that they correspond to the design parameters. Mistakes can occur when suppliers are loading the goods in their warehouses, for example, dispatching the wrong inverter, and if not checked beforehand this could lead to avoidable accidents, including equipment failure, which is expensive in both time and money.

Even quality components can fail if they are not correctly installed and maintained. The installation should only be carried out by qualified and trained personnel following all the basic safety precautions, ensuring a safe and professional installation. The corresponding equipment operation and installation

http://dx.doi.org/10.3362/9781780447810.006

88   SOLAR PUMPING FOR WATER SUPPLY

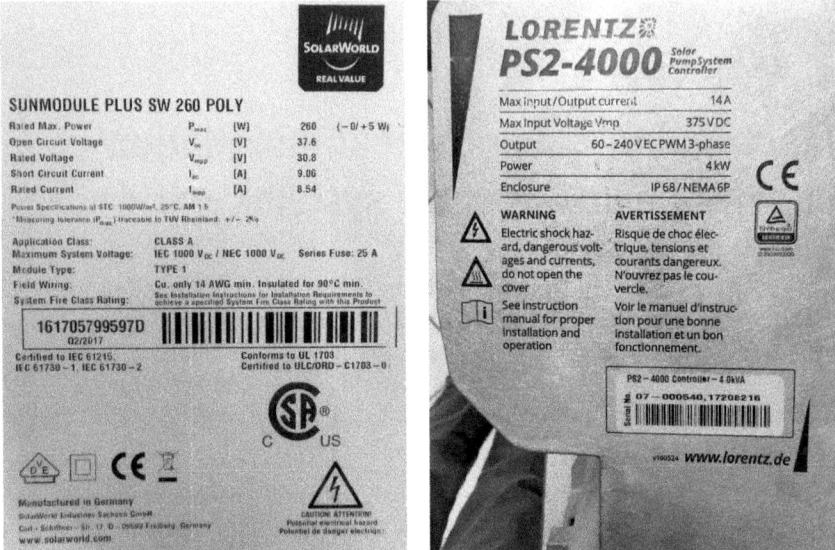

Figure 6.1 Sample nameplates for (left) PV module, and (right) solar controller
Source: Davis & Shirtliff Ltd

manual should be used. These will usually be supplied by the manufacturer together with the equipment. The instructions and recommendations in the manual should be followed strictly to avoid unnecessary mistakes that could lead to failure. Even experienced technicians who have undertaken similar installations before should be keen to read the instructions provided as manufacturers will often update their products, incorporating new instructions in the manual that experienced technicians are not aware of.

> An installation checklist developed for water technicians to monitor installation by private-sector contractors can be found in Annex J.

Reputable manufacturers provide a warranty for the equipment but this warranty can be voided if failure is caused by installation mistakes, that is, installation carried out contrary to manufacturer instructions. Besides that, installation mistakes are a safety risk to those involved in the installation through electrical shocks, which can be fatal in addition to being a hazard to the environment.

The equipment installation instructions should be used together with local/national regulations for solar-powered water systems, electrical safety codes, codes of practice, appraisal of local conditions (theft, flooding, lightning, etc.), and other institutional minimum standards from the implementing organization. As a minimum, all SPWS installations must reference IEC 60364-7-712.

Equipment failure occurring shortly after installation is often due to trivial mistakes, such as faulty cable splicing, loose cable connections, or wrong PV module wiring delivering a high voltage, leading to inverter burn out, which if careful attention was paid could have been easily avoided.

To support with supervision of installation, an installation control checklist can be found in Annex J, on the Global WASH Cluster 'Resources' webpage (GLOSWI, 2018f) or on Energypedia (2020). This checklist will help with controlling installation quality and ensuring compliance with design requirements and standards.

Finally, after installation is complete the site should be left free of clutter, all forms of debris, tools, and equipment, which can be a hazard to the operators and negatively affect the performance of the system.

6.2 Installation sequence and process

From field experience the recommended sequence of installing the pumping system is to begin with pump installation, followed by installation of all controls (disconnection switch, combiner boxes, controller) together with all cabling, and lastly installation of the support structure and the solar modules.

The installation team should be equipped with the correct tools that guarantee professional work. These tools include DC and AC clamp meter, multimeter (minimum 1,000 VDC), array tester (minimum 1,000 VDC), wire stripper, crimping tool, and compass. Optional tools include inclinometer and irradiance meter.

6.2.1 Cable splicing and dry-run protection

A splicing kit that is rated for underwater use and matched to the cable size should be used to connect the submersible cable to the motor tail cable. This splicing kit is not unique to SPWSs and is used in traditional pumping systems. The manufacturer will provide labelling or colour coding on the motor cables to allow for correct wire sequencing.

> Cable splicing is one of the most critical tasks of the entire installation. Sloppiness in workmanship can lead to an immediate or future failure.

The contracted supplier/installer should be asked to make the cable splice in their workshop and test the pump unit before delivery to site. Since damage can occur in handling and in transit, cable connections and splices should also be checked onsite before lowering the pump by conducting continuity and resistance tests on the motor and cable. It is also often possible to test small pump units in a drum/bucket of water connected to one string of unmounted modules laid out on the ground. This helps to detect any problems prior to installation and avoid the time-wasting exercise of installing a faulty pump.

Submersible pumps must be operated while completely submerged in water. Surface pumps must be operated with water in the suction pipe. A critical sensor that must be installed with every pumping unit is a dry-run protection sensor. This works by disconnecting power supply to the pump when the source water level drops below a pre-set level. Different dry-run protections are available, including well probes and water sensors/electrodes. These sensors are connected through a cable to the corresponding sensor input in the solar controller. Failure to install a dry-run protection can lead to premature failure of the pump.

6.2.2 Pump installation

Submersible pumps. Beginning with pump installation is especially important for solar retrofitted submersible pumping systems where it is necessary to remove the existing pump in order to fit a dry-run probe. It is not uncommon to find that the specifications of the existing pump are different from what had been reported during the design stage. If the pump is found to be smaller than envisioned, for example, this would call for a smaller solar PV generator and a smaller controller, which can be exchanged from the contractor.

Even for new pumping systems it is possible to encounter surprises, such as a smaller borehole casing than what was thought, necessitating a reduced size of pump, PV modules, and inverter, or a non-plumb casing that restricts the pump from entering the borehole, or a dry borehole which would mean the project has to be abandoned.

When such discrepancies are encountered, because the installation started with the pump they can be corrected relatively easily, which would not be possible if extensive PV installation works had already been undertaken.

As far as the choice of a new system versus a retrofitted system is concerned, sound judgement should be made having all the information about the system and considering the benefits to be realized. In some cases the existing pump is a new pump, thus negating the need for a new pump. In other cases, the pump is old and its soundness questionable, requiring a new pump to be bought. An important consideration in this case is to ensure that the pump to be powered with solar is adequate for variable-speed operation, is the right size for the required demand as per the design, provides an optimal design relative to the number of modules and inverter size required, does not jeopardize the equipment warranty (e.g. some suppliers may not warrant their inverters if they are installed with third-party pumps), and does not jeopardize the performance warranty of the system.

Last-minute decisions on the size of the pump relative to the well diameter have been a source of anguish (because of additional transport costs, project delays, unhappy users, and embarrassment) for many agencies that fail to give it prior consideration. Pump dimensions should be checked before procurement to ensure that both the pump (including the cable guard) and the motor can fit into the well. The minimum allowance between the pump and the well casing is usually provided by the manufacturer.

Any submersible pump installation is usually done using a hydraulic winch that lowers the pump into the well (Figure 6.2). Small pumps can be installed

ELECTRICAL AND MECHANICAL INSTALLATION OF SOLAR-POWERED   91

Figure 6.2 Pump installation using a hydraulic winch
Source: Davis & Shirtliff Ltd

into shallow wells by hand or by using a simple pulley-and-rope system. For deep boreholes, the installation work is more complex and problems such as hitting the sides of the well, abrasion of the cable, and in the worst case getting stuck or dislodging in the well, have been encountered. To avoid such problems, professional installers or people who are conversant with the process should be engaged as they are familiar with what care should be taken to ensure a well-aligned installation free of damage.

The pipe size and type used during the design stage to compute head losses must be used in the installation. If a smaller or rougher pipe is used, it will result in reduced pump performance. The submersible cable should be secured onto the pipe using plastic cable ties to prevent it from sagging and being strained.

Importantly, the position of the submersible pump in the well has an impact on the longevity of the pump. The submersible pump must be installed:

- suspended in the well to prevent contact with sand and mud at the bottom of the well;

- below the dynamic water level to avoid the pump cutting out on low water during prolonged pumping;
- above the main aquifer to ensure efficient motor cooling as the water flows around the motor;
- within a plain casing for proper cooling, to avoid turbulence on the inlet, and to prevent sand from entering the pump.

Effective circulation of water around the motor during operation is particularly important for submersible pumps as this provides the needed cooling of the motor that prolongs its life.

Submersible pumps are sometimes installed with a stainless-steel safety rope for additional safety, to avoid it falling into the well when pipes corrode or when an unexpected weakness in the piping system occurs. Plastic drop pipes must always be installed with a safety rope. The complete borehole installation should be fitted with a high-quality steel wellhead cover that is strong enough to support the complete installation.

> A cooling sleeve can be used to provide proper cooling in cases where the pump cannot be installed above the main aquifer and within a plain casing.

As a standard, submersible pumps come fitted with a non-return valve at their outlet to prevent the water in the pipe from flowing back into the well when the pump stops.

Surface pumps. Surface pumps are mounted/bolted on a concrete plinth (platform) of appropriate dimensions and strong enough to withstand the pump weight and vibration during operation (see Figure 6.3). Since surface pumps are air cooled, enough ventilation should be provided for good air circulation. The exterior of surface pumps should not get into contact with water and when installed outside should be protected from rain and direct sunlight to prolong their lifetime and reduce maintenance requirements.

Surface pumps should be installed close to the water source within the suction lift limit of the pump (see Annex A). Bends, valves, and fittings should be minimized on the suction side of the pump as they compromise the pump suction capability. Sharp bends and constrictions (e.g. pipe reductions) must be avoided too on the suction side. As a general rule, surface solar pumps should be installed as close to the water source as possible, despite their potential capacity to lift water from deeper sources. A low suction head significantly improves daily flowrate and operational reliability. A surface pump must be primed before operation, that is, the suction pipe must be full of water for the pump to operate. To prevent water from running back into the well when the pump is switched off, a foot valve or non-return (check) valve is fitted at the end of the suction pipe. The valve keeps the suction full of water. In the absence of the foot/check valve, the suction pipe will need to be manually primed (filled with water) each time the pump is started.

Figure 6.3 Horizontal surface pumps installed on a concrete plinth in Itang Water Supply, Ethiopia

To prevent cavitation in a surface pump (see section 3.3.1, 'Surface pumps vs submersible pumps') the following should be observed on the suction side:

- The maximum suction lift limit of the pump should be observed (Annex A) with reference to the net positive suction head (NPSH) requirement of the pump from the manufacturer's datasheet. The NPSH available in the pump location should always be higher than the NPSH required by the pump for cavitation to be avoided.
- Friction on the suction side should be minimized. Sharp bends should be avoided (45° elbow instead of 90° elbow, long radius instead of short radius elbow); the suction pipe should be kept short (maximum 10–20 m); the number of bends should be kept to a minimum (preferably only one bend or none at all if possible); and the suction pipe should be large enough and never smaller than the pump inlet diameter.
- The suction pipe should be kept air tight, i.e. no water leakages should be present.

Cavitation and minimum suction lift issues are not discussed in detail here as they are common to all mechanized pumps regardless of the power source. However, these two issues are highly volatile in solar pumping because power,

94 SOLAR PUMPING FOR WATER SUPPLY

pressure, and flowrate are varying constantly. The application of solar surface pumps calls for computer-aided system design, which allows the designer to analyse the behaviour of the system under all operating conditions.

6.2.3 Installation of controls

Installing controls before installing the PV modules is also recommended. This is because the controls, specifically the DC disconnection/isolation switch (see section 3.4.2), provides a way to manage the high voltage and current produced by the modules which would otherwise be impossible and dangerous to handle (see Figure 6.4).

The solar controller should be well ventilated and depending on its enclosure/protection rating, be shielded from the elements such as rain, dust, and sunlight. Some manufacturers provide controllers that are rated IP66, which means they are protected from splashes of water and dust. They should not however ever be immersed in water.

Controllers should be installed as close as possible to the PV generator to keep cable losses (voltage drop) on the DC side at a minimum.

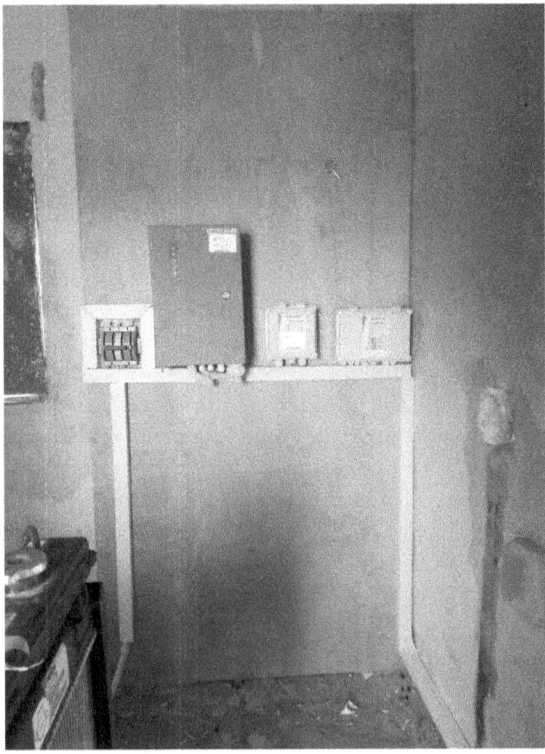

Figure 6.4 Installed controls (left to right): AC changeover switch, inverter, surge protector, PV disconnection switch

Increased distance between the controller and the motor causes harmonics that require a filter to avoid damage to the pump and the controller (see sine-wave filter in section 3.4.5). Commonly, they are installed under the solar module structure, which provides good protection against rain and direct sunlight, or in a small control room close to the solar panels, fixed at a minimum height from the ground of about 1.5 m. When installed inside a room, enough ventilation should be ensured for air circulation that will provide efficient cooling of the inverter.

Controllers must never be installed in a fully enclosed cage as this will cause them to overheat and perform non-optimally. Where theft or vandalism is a concern the inverter can be mounted inside a lockable meshed cage, as shown in Figure 6.5. The cage should also be perforated at the top and bottom to allow for an effective cooling stream of air flowing behind the controller. Refer to the manufacturer manuals in case of uncertainty.

Figure 6.5 Inverter installed in a lockable meshed enclosure under the PV array at Turkana, Kenya
Source: Oxfam

One solar pump controller is designed to power one pumping system. No other load should be connected to the solar controller except the pump. In addition, the PV generator discussed in this book is designed to power the pump alone and hence no other load should be connected to the PV generator.

6.2.4 Cabling

Cables between the pump, controls, and PV generator should be appropriately sized, as explained in Annex D. For correct cable termination at the motor and in the controller, refer to the motor and inverter manual to ensure correct sequencing. Reference should also be made specifically to IEC 60364-7-712, 60947-1, and 62253.

DC cable terminations should be carried out in such a way that it should never be necessary for an installer to work in any enclosure or situation featuring simultaneously accessible live PV string positive and negative parts.

A watertight submersible cable should be used inside the well and should be protected from physical damage especially at the point where it is in contact with the borehole casing. It should be secured onto the drop pipe using cable ties to relieve strain. The right type and size of cable splicing should be used to connect the submersible cable to the motor tail cable and should be done by an experienced technician to ensure a secure connection. The manufacturers' splicing instructions should be followed carefully. The weight of the pump should be supported on a rigid pipe (e.g. steel pipes) and not on the electrical pump cable. A separate corrosion-resistant safety rope should be used to support the pump when plastic pipes are used. The pump cable should never be used to pull the pump out of the well because this will damage the cable.

Surface cables should be approved for external use (e.g. armoured type) or for running the cable inside an electrical conduit. Surface cables should always be buried in the ground to a minimum depth of 0.5 mm for electrical safety, away from flood-prone areas. Cable glanding should be the right size to ensure a good seal, which will prevent entry of dust, insects, rodents, and moisture which can cause damage.

All connections should be made in easy-to-access junction boxes where they can be inspected, repaired, and mechanically secured. All electrical connections should be protected against water, dust, and insect intrusion.

Good cable management is important in a professional installation. It also helps with preventing physical accidents (e.g. tripping on cables) and electrical accidents (e.g. short-circuits when cables overlap). See Figures 6.6 and 6.7 for examples of poor and good cable management.

6.2.5 Module mounting structures

Different types of module mounting structure are available commercially or can be fabricated locally. Ground mount and pole mounts are the most

ELECTRICAL AND MECHANICAL INSTALLATION OF SOLAR-POWERED   97

Figure 6.6 Examples of poor cable management in South Sudan

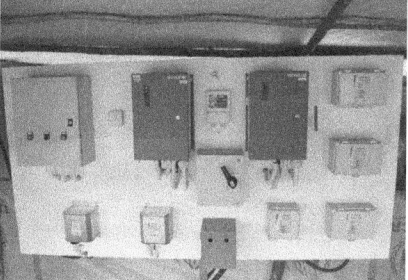

Figure 6.7 Examples of good cable management in South Sudan and Tanzania respectively

common structures for off-grid applications, especially in rural contexts, where most humanitarian and development work is done.

Ground mount. The structure is directly anchored onto the ground either in cast concrete or bolted onto a reinforced concrete block. It is the most robust configuration as it is mounted not more than 1 m from the ground and hence is less affected by wind. Orientation and tilt can also be optimized. Ground mounts have the advantage of easy access for cleaning but expose the modules to theft and vandalism. (See Figure 6.8 and Figure 5.9 for an example of a ground mount structure.)

Pole mount. These are anchored into the ground on a high support pole (usually more than 1.5 m high on the lower side) and filled with reinforced concrete (See Figure 6.9). This structure is most affected by wind loading

Figure 6.8 Ground mount structure in Kawrgosk refugee camp, Iraq

Figure 6.9 Pole mounted PV modules in Bidibidi refugee settlement, Uganda

which should be factored in when designing the structural member sizes. Pole mounts provide security against theft and vandalism but are more expensive and more difficult to access for cleaning. This configuration is adaptable as it can be constructed on almost any terrain, orientation and tilt can be optimized, and modules can be raised high as an anti-theft measure. Cleaning can be difficult as the modules are often raised high above the

ELECTRICAL AND MECHANICAL INSTALLATION OF SOLAR-POWERED 99

Figure 6.10 Roof mount on a flat tank roof in Gaza

ground. Due to the wind effect, this structure is most prone to failure if it is not correctly designed and installed. Adequate bracing should be provided to prevent swaying. The concrete for the pole support foundation should be mixed and allowed to cure properly to provide adequate support against wind and rain. Most failed structures encountered in the field are pole mount structures.

Roof mount. The structure is mounted on an existing roof, either sloping or flat, often in the same direction and tilt as the roof (see Figure 6.10). A roof mount has the advantage of requiring less land (or no extra land at all if the existing roof is enough) but on the flipside its feasibility depends on the roof and it is not always possible to optimize the orientation/inclination, resulting in energy losses which must be factored into the solar power calculations. Roof mounts are difficult to access for cleaning.

Ground screw. This structure is a relatively new concept which does not require excavation, concreting, or refilling. The support poles (ground screws) are simply hammered into the ground, resulting in minimal disturbance on the soil and vegetation (See Figure 6.12). They are advantageous in the sense that they are easy to install, easy to remove, and reusable; however, they can only be installed on stable soil types and in non-rocky areas.

Elevated tank mounting. Sometimes modules are mounted on an elevated tank for security, reduced cost, and efficient space utilization (See Figure 6.13).

Solar tracking systems are discussed in section 7.3 and are installed only on pole mounts, which allow the modules to be manoeuvred to follow the daily and seasonal path of the sun in order to maximize the energy yield.

Figure 6.11 Combined roof and pole mount installation at IOM water scheme, Kutupalong Balukhali Expansion Site refugee camp (Cover photo)
Source: IOM-Bangladesh

Trackers can be a single axis (seasonal changes) or two axes (daily and seasonal). Trackers were previously common in higher altitude locations and less common in the tropics, where they have become less popular due to the affordability of PV modules, such that the benefit attained by installing trackers can now be realized more economically by oversizing the PV generator. Trackers are also undesirable as they are expensive and require maintenance, which is difficult in remote locations.

Factors to be considered in the structure installation include:

- *Location.* The tilt angle of the structure should be set equal to the latitude of the location (with 15 degrees being the minimum tilt angle for self-cleaning of the modules when it rains) and facing the equator. In instances where there is great risk of modules being blown off by strong winds the tilt angle should favour module survival under strong winds.
- *Shading.* The structure should be installed away from any shadowing, such as that cast by nearby trees, buildings, or overhead cables.
- *Soil conditions.* The soil should be firm enough to carry the weight of the whole system. Foundation depth and concrete casting should be sufficient to provide support according to the soil type.
- *Rain.* For pole mount and ground mount structures the metallic support poles should be cast in concrete to a height above the flood level to prevent corrosion due to flood waters.
- *Temperature.* Spacing between the modules should accommodate the expansion and movement of the modules with changing temperature (see section 6.2.6).

ELECTRICAL AND MECHANICAL INSTALLATION OF SOLAR-POWERED 101

Figure 6.12 Example of ground screw structure

Figure 6.13 Tank-mounted structure in an IDP camp near Maiduguri, Nigeria

102 SOLAR PUMPING FOR WATER SUPPLY

- *Corrosion.* Anti-corrosion measures, especially for mild steel structures, should be put in place, and stainless-steel hooks and bolts should be used to prevent corrosion at the contact point. The structure should be properly coated and protected against environmental factors, such as rain, humidity,and salinity, as failure to do so will affect its structural integrity.

Assessments done in several countries in equatorial regions have found that mounting structures, if not properly designed and installed, are prone to failure. Such failures (tilting or complete tipping) can happen under normal loading if inadequate foundation depth (especially for the pole-mount structure) is provided (Figure 6.14). To reduce the likelihood of such disastrous failures the foundation must be designed to carry the expected point and wind loads, as determined using the relevant structural design standards and codes.

6.2.6 Solar module installation

Mounting the solar modules onto the support structure is the last step of the mechanical installation of the PV system. All DC controls and wiring should be completed prior to this step. This will allow effective electrical isolation of the DC system (via the DC disconnection switch and PV module cable connectors) while the array is being installed, and effective electrical isolation of the PV array while the inverter is installed.

Solar modules should be mounted onto the support rail and then firmly fastened using bolts, j-hooks or one-way screws through the mounting holes. Clamps can also be used but must not come into contact with the front glass of the module or shadow it. Modules must be installed in the clear open without any shading above. A minimum space of about 10 mm should be kept between modules to allow for expansion and aeration. The module must be fastened on all four corners to prevent it from being blown off the support due to inadequate bolting (see Figure 6.14).

Once all the modules are firmly secured on the structure, the home-run negative and positive terminals from the PV array are connected to the DC disconnection switch and/or combiner box but omitting the module interconnections. Note that while the installer will be handling live cables during the subsequent module interconnection, because the circuit is broken at the DC disconnection switch, there is no possibility of an electric shock current flowing from the partially completed PV string. The maximum electric shock voltage that should ever be encountered is that of one individual PV module.

The interconnection is done using the male/female cable terminals at the back which must be pushed in all the way to make a tight connection (see Figure 6.15). The connector can be MC3 or MC4 type although the former

Figure 6.14 Failed pole mount in Yida, South Sudan (left), modules blown off in Fafen, Ethiopia (right)

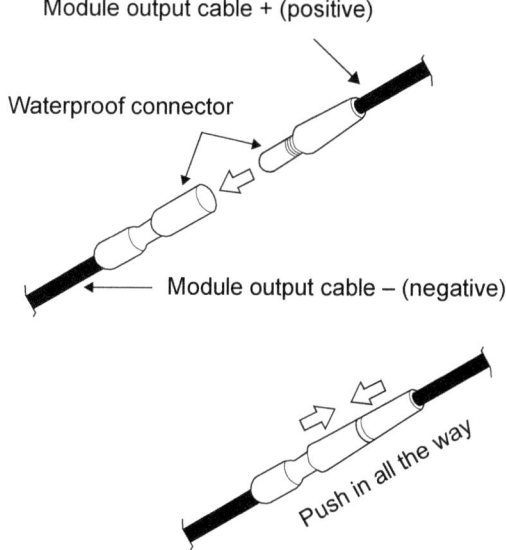

Figure 6.15 A tight connection of solar module MC3 quick connectors

is mostly obsolete. Before connecting modules, check the contacts are corrosion free, clean, and dry. Modules are connected in series to increase the operating voltage by plugging the positive plug of one module into the negative socket of the next (discussed in section 5.3.5). The system voltage is defined by the DC controller or inverter, as discussed in Annex B, but must be less than the maximum system voltage allowable for the module, which is typically 1,000 V. Some recent modules have a maximum system voltage of 1,500 V.

104 SOLAR PUMPING FOR WATER SUPPLY

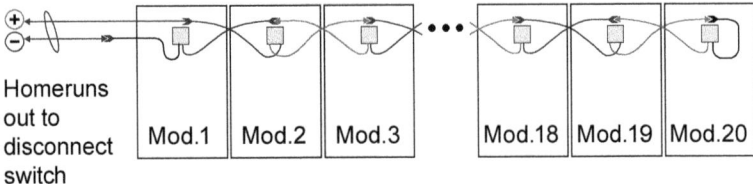

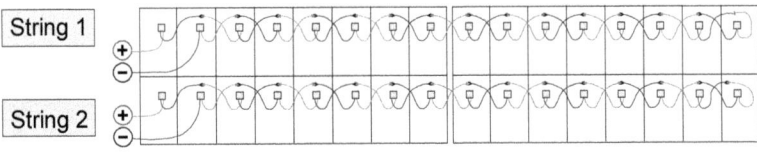

Figure 6.16 Leap-frog connection of modules

The voltage of each individual string should be verified before making a parallel connection. A difference of more than 10 V between strings should be checked and rectified before making the parallel connection.

The interconnection adopts the leap-frog style of connecting modules whereby one terminal of the first module goes into one terminal of the third module, while the remaining terminal of the first module goes into one terminal of the second module. The remaining terminal of the second module goes into the first terminal of the fourth module, while the remaining terminal of the third module goes into one terminal of the fifth module. This is repeated until the end of the string (See Figure 6.16). The purpose of this leap-frog connection is to achieve a free positive and free negative terminal (home-run terminals) on the same side of the string as opposed to having the two terminals on opposite sides, thereby reducing the DC cable and distance required to run to the disconnection box.

The most optimal sequencing should be adopted depending on the location of the controls, for example, if the inverter is positioned midway under the panels then the connections will be done so that the positive and negative terminals meet in the middle of the array.

During installation of solar modules, it should be ensured that:

- no module is damaged, e.g. the module covering glass is not broken, the cable terminals have no cuts or scratches;
- all the modules are of the same rating (same model, current, voltage, and power);
- each string should have an equal number of modules;
- the voltage in each string is the same. This helps to show if there are any module wiring mistakes and to ensure the right voltage goes to the inverter;
- the terminals cables are not hanging or caught on plants but are neatly secured using plastic cable ties onto the metallic structure. The cables

ELECTRICAL AND MECHANICAL INSTALLATION OF SOLAR-POWERED   105

Figure 6.17 Example of untidy and tidy module cables

should not touch the back of the modules as they can overheat and melt (see Figure 6.17);
- the positive and negative home-runs are clearly labelled with positive and negative markings, or by using colour coding;
- the angle bars that hold the modules should be installed in such a way that they do not obstruct or cast a shadow on the solar cells.

A maintenance walkway should be provided for cleaning and other maintenance activities such as module replacement and cable inspection (see figures 6.18 and 6.19 for examples). This is especially important for dusty environments and locations where self-cleaning cannot be guaranteed.

Figure 6.18 Walkway on a roof mount at Strathmore University, Kenya

Figure 6.19 Walkway on a ground mount in Koboko, Uganda

6.3 Earthing, lightning protection, and surge protection

High voltage peaks induced by lightning are one of the most common causes of electronic failures in outdoor installations. Lightning surges are also a common cause of controller failure in SPWSs.

The probability of lightning striking a PV installation depends on the size and location of the system as well as prevalence of thunderstorms in the area. Installation of a PV system does not increase the likelihood of a lightning strike.

A PV system can produce large amounts of current and voltage that can be dangerous if not well handled and managed. To arrest such dangers and ensure the safety of the system all exposed or accessible PV equipment and circuits must be properly connected to earth (grounded). Properly grounding PV systems reduces the risk of electrical shock to personnel and the effects of lightning and surges on equipment.

Earthing and grounding should be referenced to IEC 60347-7-712 and 62548 as well as complying with the requirements of the country electrical codes. Briefly, the general considerations for protecting PV systems from surges are:

- *Lightning arrestor.* A lightning arrestor should be installed at the highest point on the SPWS site and be connected to a conductive 2.5 m-long,

16 mm copper grounding rod safely connected to the ground using an insulated copper cable. Lighting arrestors should not be installed in physical contact with the PV module structure.
- *Bonding.* All exposed metal components (including PV module frames, metallic mounting structures, metallic enclosures), ground terminals of the disconnection switch, and the controller should be interconnected using a grounding conductor, such as a copper wire of minimum size 6 mm^2 (AWG#8), and the conductor connected to a protective earth connection.
- *Grounding.* One or more 2.5 m-long, 16 mm conductive copper grounding rods should be driven into the ground and one end connected to the PV array and controller using a cable of sufficient size. The rod should be driven into sufficient depth that most of its area contacts conductive soil (moist soil), so that when a surge comes down the line, the electrons can drain into the ground with minimal resistance. Low earth contact resistance can be improved by adding soda ash or activated charcoal to the area surrounding the rod.

A properly grounded conductive structure will guide the lightning surge or voltage spike into the earth (divert the surge) and reduce the potential for damage.

Grounding is a complex issue that should always be referenced to manufacturer installation manuals as well as other, more detailed documents, such as the Solar America Board for Codes and Standards (SABCS, 2015).

Surge protection devices (SPDs) provide additional protection for electronic equipment (see Figure 6.20). They are used to reduce the risk of lightning damage entering from the cables to the equipment. Though SPDs greatly reduce the risk of damage, they do not eliminate it completely. It is recommended that one SPD unit is installed at every sensor input in a PV pumping system. Sensor inputs include dry-run input, low/high level input, and pressure switch inputs.

Figure 6.20 Lorentz surge protection device
Source: Lorentz

In addition, the probability of damage caused by far-range lightning strikes can be minimized by keeping the cable distance between the PV array and the controller short, as well as by burying all exposed wires into the ground instead of running them overhead.

6.4 Electrical safety

Electrical safety when installing a PV pumping system should be given proper attention by minimizing potential hazards and risks. That said, long-term safety of the system begins at the point of correct design followed by proper installation, operation, and maintenance.

As PV power cannot be switched off, a means of isolating the power before attempting to work on the PV pumping system should be provided (as a minimum, a DC switch of the right rating will always be present between the PV modules and the inverter). Failure to take precaution can lead to injuries to personnel and damage to equipment.

Labelling of DC cables is important to make it easy to distinguish between different pairs of negative and positive circuits. Danger signs should also be put up at points where high voltage is a risk. All labels must be clear, easily visible, and constructed and affixed to last and remain legible for the lifetime of the system.

Additionally when solar modules are installed on poles or rooftops, working at heights poses risks for the installers and operators of the system. Appropriate measures should be taken to protect human life, such as through training and mitigating against the risk of falling.

The most important safety measure in any PV installation is the human factor which, if neglected, can lead to avoidable failures and injuries. This requires that only personnel who are acquainted with the voltages present in a PV pumping system be allowed to work on any live components of the system (see more at section 11.2.6).

CHAPTER 7
Specific considerations and limitations for solar-powered water pumping

This chapter analyses a number of issues specific to solar pumping which often lead to misconceptions and design errors. These relate to variable chlorination, solar tracking pumping in hot water contexts, tank sizing, theft-avoiding measures, and overpumping related to the use of solar energy sources. An example of solar emergency kits and a selection of some of the most frequently asked questions received on the solar helpline run by the authors is shown at the end of the chapter.

Keywords: solar tracking, water chlorination, vandalism, theft, solar pumping schemes, dosing pumps, hot water pumping, aquifer overpumping, solar emergency kits

7.1 Chlorination

Water treatment with chlorine or chlorine compounds is a usual measure taken in humanitarian operations to ensure bacteriological safety of drinking water. Water chlorination entails the addition of a fixed amount of chlorine per volume of water to be treated. The amount of chlorine to add is determined based on the quality and quantity of the water to be treated (see more on chlorination in Skinner, 2001).

When water is pumped at a fixed rate, as happens when the water pump is connected to a constant and stable energy source, such as a diesel generator or a stable grid, it is easy to calculate how much chlorine needs to be added in order to achieve bacteriological safety.

In solar pumping schemes, the solar radiation that hits the panels will change in value several times per second. As a consequence, the water flowrate delivered (m^3/hr) will change all through the day. Therefore, in solar-powered water schemes, the fact that water flowrate varies during the day and daily water pumped varies during the year needs to be taken into account.

There are four different solutions commonly found in the field in order to solve this problem: tank chlorination, in-line mechanical dosers, variable dosing pumps, and valve-regulated chlorine dosers.

7.1.1 Tank chlorination

The chlorine dosing is done at the water tank. Since the water volume of a full tank is known, it is easy to calculate the amount of chlorine to be added.

http://dx.doi.org/10.3362/9781780447810.007

110 SOLAR PUMPING FOR WATER SUPPLY

Tank chlorination is a basic approach yet it does require more constant monitoring and additional human resources.

7.1.2 In-line mechanical dosers (water powered)

A device is installed in the pipeline between the pump and the water tank which basically consists of a valve which opens at intervals, dosing in a fixed amount of chlorine that can be manually adjusted. The higher the water flowrate, the more times the valve will be opened, releasing a higher amount of chlorine, and vice versa, so the chlorination is proportional to the water flow. An operator will need to regularly refill the doser with chlorine.

While these devices are easy to install and use, reports from the field indicate that they tend to get stuck at the seal level from time to time. Mixing the chlorine (fully dissolved granules) and positioning the screen of the pickup hose so it does not touch the bottom of the chlorine container ensures longer, trouble-free use.

7.1.3 Variable dosing pumps (electric powered)

There are a whole range of chemical dosing pumps in which the chemical dose rate (e.g. chlorine) is varied in direct proportion to changes in flow, measured with a water flowmeter. This method of control maintains a chlorine/water ratio constant (mg/l) and does not make any allowances for variations in the incoming water quality (see more at Brandt et al., 2017).

While an operator will be needed to start and stop the pump and refill the chlorine tank, as for in-line water-powered chlorinators, an extra inconvenience comes from the fact that an electrical power supply will be needed for the dosing pump to function. As typically solar water pumping is used where no reliable grid is present, an alternative energy source (typically a small panel and battery, since the energy demand for these pumps is normally very low) will need to be provided to ensure proper functioning. Note that in case of hybrid systems and water pump functioning beyond the solar day, another power source for the dosing pump will be needed to ensure proper water treatment.

7.1.4 Valve-regulated chlorine doser (manually operated)

More traditional solutions found in the field consist of a parallel pipe connection conducting part of the water flow through a chlorine container (see Figure 7.1). A valve mounted on the pipe going through the chlorine container is manually regulated to limit the flow of water and therefore the amount of chlorine going to the main pipeline. After some iteration and careful monitoring of chlorine residual, an experienced operator will be able to know whether the valve should be almost closed (cloudy days with lower water flows) or more open (for sunnier days).

Variations are systems using tablets and/or a venturi restriction (see more in Skinner, 2001, pp. 22–24) on the pumping main, or on a bypass to it that

SPECIFIC CONSIDERATIONS AND LIMITATIONS FOR SOLAR-POWERED 111

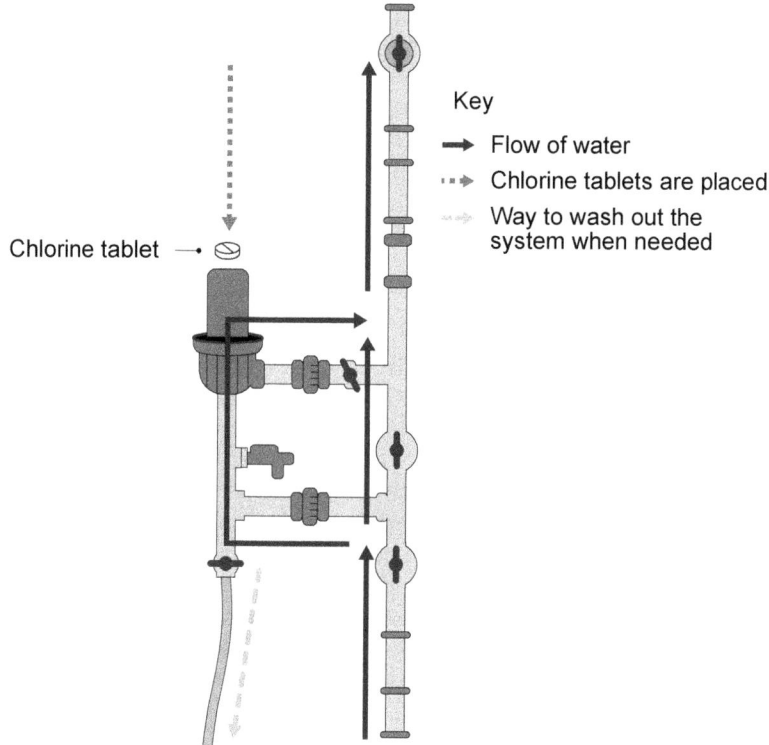

Figure 7.1 Valve-regulated bypass chlorinator

takes some of the flow so that a small hose at the venturi can be used to suck the solution into the pipe. Valves can be used to adjust the flowrate in the bypass pipe and/or in the small pipe to get the right dose.

7.2 Full-tank detection at long distances

Pressure switches are control devices that open or close electrical contacts in response to changes in pressure. Float switches are mechanically actuated devices for water-level detection.

In applications where water tanks are just a few metres away from the water point, a float switch can be easily installed within the tank to detect water levels. When the water tank is full, an electrical signal is sent via a cable to the controller of the pump to stop pumping and vice versa.

However, when the water tank is hundreds of metres away from the water point this configuration becomes impractical since such a long cable would be too expensive and could potentially, in some contexts, attract lightning, which would put the whole water scheme in danger.

A more practical solution for these contexts is to install a pressure switch in the pipeline together with a simple ball valve in the tank. As the water tanks get full, the ball valve starts closing and pressure increases in the pipeline. When the pressure reaches a certain pre-established threshold, the pressure switch sends an electrical signal to the pump controller (located just a few metres away) to stop the pump. The system switches back on automatically when the switch-on pressure is reached. With this configuration, there is no cable from the ball valve to the pump controller, which avoids long cables and the problems that come with it.

7.3 Solar tracking

Because of the different position of the sun in the sky during the day and through the seasons, the only way to have solar panels at a right angle with the sun at all times, and therefore maximize electricity production and water output, is to move the panels to follow the sun's path (see Figure 7.2). This is the function of mechanical tracking devices installed in a solar array. A tracking device can increase electricity output up to around 30 per cent.

As mentioned in section 5.2, while tracking devices have been used extensively in the past they are used infrequently now in solar pumping applications, for reasons of both high cost and complexity to maintain and repair versus the more economical and simpler option of adding more solar panels to the array. A tracking device is therefore discouraged, especially in humanitarian operations and in places where operation and maintenance of these trackers cannot be guaranteed.

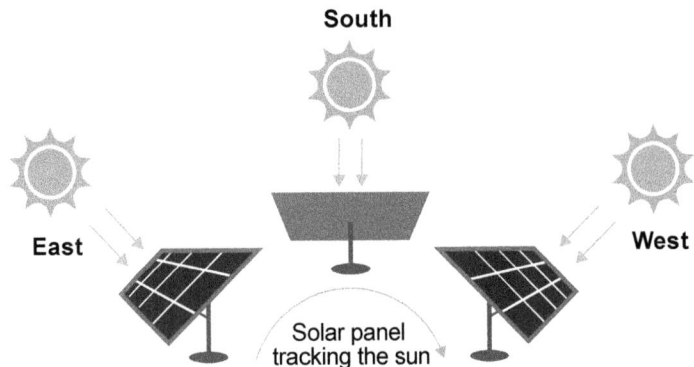

Figure 7.2 Solar trackers: double-axis tracking (left) and single-axis tracking (right)

7.4 Emergency solar pumping kits

Emergency solar pumping kits are modular, making them quick and easy to install at any moment, which is especially useful in contexts where fuel supply is limited or irregular, the grid is unstable or non-existent, and emergency water supply systems need to be installed (so no time for normal procurement procedures).

SPECIFIC CONSIDERATIONS AND LIMITATIONS FOR SOLAR-POWERED   113

Table 7.1 Summary of Global Solar and Water Initiative specifications for small, medium-sized, and large solar pumping kits for rapid deployment

Kit	Flow range (m^3/h)	Daily flow range (m^3/day)	TDH range (m)	Lorentz pump model	Grundfos pump model	Solar panels (W)
Small	1.0–2.8	10–30	40–120	PS1800 HR-14H 1.7 kW pump	SQF 2.5-2 1. 4kW pump	615–4100
Medium	3.5–7.0	30–70	60–140	PS4000 C-SJ5-25 4 kW pump	SP5A-33 3 kW pump	2460–9840
Large	6.0–12.0	50–100	80–180	PSk2-9 C-sj8-44 7.5 kW pump	SP9-32 7.5 kW pump	6560–15580

TDH = total dynamic head

A few WASH agencies with in-house technical expertise have developed their own kits. While it is strongly recommended to use quality design software to get the right dimensioning of the solar system for each individual water system, the Global Solar and Water Initiative team has developed guides for small, medium-sized, and large emergency solar pumping kits, with technical specifications (GLOSWI, 2018e).

This provides a quick reference for emergency deployment. The performance ranges indicated on each kit are meant to be a rough guide and should not be taken to be the actual installed performance. These estimated performance ranges are summarized in Table 7.1.

The terms small, medium, and large do not refer to the solar water pumping system sizes available in the industry, but rather to the emergency kit sizes relative to each other.

These kit sizes are based on what was most widely encountered in field assessments of 160 water schemes in 55 refugee camps and communities, mostly located in Sub-Saharan Africa. Smaller and larger solar water pumping solutions are available from different manufacturers. Off-the-shelf solutions of up to 75 kW (motor size) are available from some good-quality manufacturers.

The performance estimations of each kit (shown in Table 7.1) are based on the Lorentz and Grundfos sizing software. These software programs use long-term estimates of solar and meteorological data from accredited sources, such as NASA.

7.5 Power range for pump motors and inverters

There are widespread misconceptions about the limitations of pumps powered with solar energy in terms of water volume and depth of boreholes (or total dynamic heads – TDH).

In the past, the term 'solar pumps' was used to describe water pumps that could be directly coupled with solar panels. Those solar pumps were equipped with DC motors, which for a number of reasons were limited in most cases to 4 kW of pump motor power. From this came the association of 'solar pumps' with small pumps.

Technical developments in inverter technology have allowed solar panels to power AC pumps with higher motor power ratings, making it possible to pump higher water volumes from larger TDHs. The higher the input power of the inverter, the bigger the pump motor power can be. Manufacturers of inverters specifically designed for water pumping have been increasing the sizes of inverters offered in recent years.

Several reputable solar pumping inverter manufacturers now offer solutions for water pump motor powers up to 37 kW. In this range, submersible centrifugal pumps can deliver water volumes of up to 240 m^3/h and TDH of up to 200 m.

For helicoidal water pumps, while the water output is quite limited to a few cubic metres per day, TDHs are up to 450 m, making it possible to pump small volumes from deep water. For surface pumps in the same range of motor size, volumes of up to 2,000 m^3/day are easily possible.

Today, larger fit-for-purpose inverters (standard inverters with in-built software designed for solar water pumping) are available off the shelf, allowing pumps from 75 kW (e.g. SolarTech, Lorentz) up to 280 kW (Fujielectric) of motor power to function with solar power (with hydraulic potential (flowxTDH) from 15,000 m^4/h up to 50,000 m^4/h).

Moreover, since inverters for large solar power plants for electricity production exist in the market, some solar pump manufacturers and private contractors offer the possibility to use these utility-scale inverters for solar water pumping applications. For example, WellPumps manufacturer uses standard Schneider inverters with its own solar water pumping software, allowing up to 220 kW of motor power.

The largest solar water pumping scheme of such kind visited by the Global Solar and Water Initiative team was found in Lebanon, using a pump motor power of 150 kW.

While this is not a complete list of brands and manufacturers of various inverter types and sizes, what is clear is that the use of the already existing inverters on the market make it possible to literally solarize any water scheme found in humanitarian operations today.

On the other end of the spectrum, pumps as small as 150 W of motor power that work with a single solar panel are now also common on the market.

7.6 Vandalism and theft

Theft of PV modules has become less common in recent years due to their low cost. Yet, as the majority of PV pumping systems for humanitarian and development work are implemented in rural, marginalized locations, solar modules are attractive commodities because they can meet the energy needs of communities. Theft and vandalism of solar modules can be a concern where:

- there is a lack of ownership;
- systems are installed at remote locations and near public places or roads;
- users have not been consulted and the energy system installed does not meet their needs;
- there is disparity in water provision with some communities feeling underserved, leading to sabotage of the others' water systems;
- systems fall into disrepair (and the energy service ceases).

Certain measures can be taken to deter or prevent theft and vandalism including:

- welding the bolts onto the structure, making it difficult to remove the panels;
- using pop rivets to fasten the modules onto the structure which require special tools to remove that will not be readily available to potential thieves;
- using elevated pole-mount structures that are hard to reach or mounting the modules on an elevated water tank (see Figure 7.3);
- fencing the perimeter of the PV module area and securing it with a lock; installing barbed wire/razor wire; installing an electric fence; fitting a motion-sensitive siren; and/or lighting the PV module area (see Figure 7.4);
- installing the PV modules close to populated areas, such as near the users;

Figure 7.3 Fenced SPWS with a solar light for security in Turkana, Kenya

116 SOLAR PUMPING FOR WATER SUPPLY

Figure 7.4 Extreme module security measures in Burao, Somalia

- staffing the PV module area with a guard, especially at night; having someone living at the water point; or having a vigilance plan where community members take turns to guard the system;
- painting the name of the village/owner on the back of the solar modules with non-removable paint;
- raising awareness among the users on the value of protecting and keeping the system safe (community ownership, see Figure 7.5);
- engaging security forces (such as police, army) in extreme security situations.

It has been observed that communities that are well aware of the value of keeping the system working will pull together to protect it, including taking disciplinary action against perpetrators, reporting to authorities, and charging the cost of repair to the families of people who damage the modules.

Preventing theft and vandalism cannot always be achieved, especially for locations that are generally insecure due to conflict or war. Consider such security concerns at initial design stage as PV water pumping may not be suitable and should be discouraged in such circumstances (see video from World Bank, 2017).

Figure 7.5 Awareness raising of the value of the SPWS to avoid theft and vandalism

7.7 Overpumping of aquifers

While the authors could not find rigorous studies showing the relationship between solar pumping adoption and overextraction of water there is a genuine concern that solar pumping could lead to water wastage and overpumping of water sources.

This is mostly based in the perception by many communities of users (but common also among implementing organizations and other water stakeholders) that, where water can be supplied by solar-powered schemes, water is free. The absence of rigorous groundwater assessments looking at the available water resource (i.e. storage plus recharge) with clear and enforced recommendations for water extraction, especially in places where an increasing number of new boreholes are being drilled and/or retrofitted to solar (see more on this in section 5.3.2), may lead to overexploitation of the available water resource.

While solar-powered water schemes will typically suffer fewer breakdowns and have much less intensive maintenance than generator or handpump schemes, solar systems can and will experience problems at some point. The absence of a plan to have funds available for the maintenance of the system will undoubtedly lead to systems falling into disrepair and abandonment. It is therefore critical that communities of users understand the limitations of solar pumping solutions so that, wherever possible, water is provided at a fee, even if supplied through solar pumping. This will be an important measure to ensure greater sustainability but also the best way to limit the feared wastage and/or overextraction of water.

Another way to limit overextraction of water, especially relevant for water schemes with multiple uses (e.g. water-for-people and water-for-food), is to transform the economy to low water use or/and to support actions that will lead to a switch from excess water use to a low water agro-economy (see for example Delegation of the European Union to Pakistan, 2018).

As the solar pump is in many cases driven by a microprocessor, it is possible to limit the daily flow to a certain agreed amount to be set at the control box/inverter level.

In addition, new liquid level sensors (vs simple dry-run sensors) measure water levels in the well (static and pumping levels) and record the data in the pump controller, allowing for proper and simpler follow-up of water levels and tracking of seasonal and long-term well behaviour to better support monitoring of water resource use. As the data recording is, depending on the manufacturer, done in the pump controller (and therefore can be accessed remotely from a central office), this is a logical addition to solar pumps – especially at critical water points – at minimal incremental cost.

Finally, there are situations where excess electricity produced can be used as a remunerative crop or for purposes other than water extraction (e.g. electricity sold via the grid; see more in Shah et al., 2018), which in turn will incentivize the rational use of solar pumping, limiting overpumping of aquifers.

7.8 Hot climate zones and hot water pumping

7.8.1 Solar panel considerations for hot climate zones

Contrary to what intuition would dictate, there are important challenges when using solar PV panels in areas with high levels of sun irradiation and increased air temperatures. Solar panels are normally tested and rated for cell temperatures of 25 °C (STC). The temperature of the cell can reach 25 °C with ambient temperatures well below that value. In real conditions, and especially in many places within the sun belt, ambient temperatures will easily reach values over 20 °C, 30 °C and even 40 °C during the day, meaning that cell temperatures will increase over the 25 °C at which the panels were tested and will start experiencing losses, with voltage output decreasing and peak energy output dropping as cell temperature rises (see sections 2.10 and 4.2).

The loss of energy output due to temperature is expressed by the temperature coefficient of the panel, which should appear in the datasheet of the product and depends on the kind of panel (mono or polycrystalline, amorphous, thin film) and the manufacturer. If, for example, the temperature coefficient is –0.5 per cent per degree Celsius, this would mean that for every 1 °C over 25 °C of cell temperature, the energy output of the panel will be reduced by 0.5 per cent. The higher the ambient temperature, the more losses induced. At an ambient temperature of 30 °C, it would not be rare to find panels at around 60 °C.

The way to compensate these losses (which are computed and accounted for in good-quality solar pumping design software) is to oversize the solar generator by adding a few more panels. In order to minimize this oversizing, it is important to try to minimize losses due to temperature. Two measures can be typically taken to do so:

1. Allow space at the back of the panels and between panels for air to circulate, cooling the panels down.
2. Chose panels with a lower temperature coefficient (typically monocrystalline will have slightly lower temperature coefficients than polycrystalline, and thin film will have the lowest of all).

7.8.2 Inverter considerations for hot climate zones

Reputed manufacturer inverters will be able to operate within temperatures of up to 60°C. However, as for solar panels, an inverter's AC output reduces as temperature rises. Starting from the rated continuous AC output at 25°C, the continuous output of inverters decreases, for example, at 40°C by 6–15 per cent, depending on the model.

In the same manner as with solar panels, a way to mitigate performance losses at high air temperatures is to oversize the inverter. The indications on the datasheets of inverter models are to be referred to for calculating how much oversizing of the inverter's capacity would be required to provide sufficient AC output at consistently elevated air temperatures.

Furthermore, the technical room where the inverter is located should have enough air inlets to ensure maximum air flow to cool down the system. Depending on the size of the solar PV system, a ventilating system may be considered to support the air flow in the technical room and keep temperatures in check (see more at Energypedia, 2019). Finally, where inverters are mounted in a casing inside a control room it is considered good practice to leave the door cover of the enclosure open during working hours, allowing for cooling and air flow.

7.8.3 Hot water pumping

Submersible pumps may be affected in hot working conditions, either underperforming or having their working life shortened when the water to be pumped surpasses a certain temperature (e.g. in areas with volcanic activity).

For helical rotor pumps, hot water may damage the rubber stator; for other centrifugal submersible pumps, it may be rubber parts that are affected. In addition, the motor in the borehole may have insufficient cooling, which could compromise the whole water scheme.

Some manufacturers, such as Grundfos, in order to protect pump and motor rubber parts, offer the possibility that pumps be fitted with bearings made of FKM material, which is resistant to liquid temperatures of up to 90°C (see Grundfos, 2012). Other manufacturers will clearly indicate the maximum water temperature at which the pump is expected to work (30°C to 40°C is a typical threshold for many pumps), and may offer alternative models for temperatures above 30°C or 40°C.

As for insufficient cooling, similar to inverters and solar panels, oversizing the pump is a way to compensate for efficiency losses or, in other words, derating the motor power. Manufacturers will provide derating factors. What is important to keep in mind is that the solar generator will have to be reviewed, and probably oversized, in order to power a bigger pump.

Example 7.1 Derating motor power for hot water pumping

A pumping scheme to provide 50 m³/day is designed for a village. It is estimated that a pump with a motor power (P2) of 4 kW will suffice to provide the required water. The water temperature is checked and found to be 40°C, while the pump temperature threshold is 30°C. At 40°C water temperature and for that size of motor, the manufacturer provides a derating factor of 0.9. What motor size will be needed in order to provide the expected amount of water?

The 4 kW motor will produce 4 kW x 0.9 = 3.6 kW of power. In order to have the expected 4 kW, the motor power is divided by the derating factor: 4 kW/0.9 = 4.4 kW. The nearest higher motor size available in the market is 5.5 kW, so a pump of 5.5 kW will have to be installed. The solar generator will have to be resized for that power.

7.9 Frequently asked questions

Can I precisely predict water outputs in a solar pumping scheme since solar conditions are so variable?
You cannot predict exactly what the water output will be in any given day. Typically, the most reputed solar design software will provide *average* water output per day for different months of the year (e.g. 30 m³/day during the month of June). Based on this, your location, and your experience, you will be able to predict a range of values per day quite accurately.

For example, you may not be able to ascertain that a certain solar pumping scheme will provide 35 m³/day exactly, but you may find that it will provide between 24 and 42 m³/day. If your needs for that particular location are below 24 m³/day, you can be certain your solar system will meet those needs.

Can I get enough water during cloudy days and in the rainy season?
Normally, if the system has been rightly sized, you will be able to get water on rainy or cloudy days too. The questions is how much water you can get and whether that will be enough to meet your needs. As explained in the question above, you can predict a range of values for water output and whether your water needs will be met. Typically, for systems intended to provide water for drinking purposes, the size of the solar arrays will be designed for the worst meteorological conditions of the year in order to ensure water needs are met even during cloudy days and rainy seasons.

Is the capital cost of solar still prohibitive, especially when compared to equivalent diesel-based solutions?

This used to be the case. However the price of solar panels has dropped 80 per cent in the last 10 years, making the technology more affordable. Typically for small schemes, the solar capital cost can be compared to that for diesel generators. What is important, however, is to look at costs over the life of the equipment. As good solar panels have a lifespan of 25 years and the running costs are so low, the cost advantages are clear when looking at it over time. On average, the return on investment period ranges between 0 and 4 years, and cost savings are of 40–90 per cent when compared to equivalent diesel systems (see more in Chapter 9).

Do I need to hire an electrical engineer to deal with my solar pumping project?

Although an electrical engineer with knowledge in solar technology would give you an additional advantage it is not strictly needed. A water engineer with a few days of training could get the necessary knowledge to deal with the main issues of a solar pumping project. Knowledgeable private-sector contractors will most likely be present in capital cities, meaning your water technician will have to monitor only some critical steps at each stage of the project.

Is there any point thinking about solar pumping in countries where fuel prices are low?

The economic case for solar may still be made (depending on national solar market conditions) since the running costs of solar pumping schemes are so low. The Global Solar and Water Initiative has run several economic analyses to compare solar with diesel systems in Sudan and Iraq (where diesel can be as cheap as US$0.3 /litre), finding that solar pumping solutions made economic sense in most cases. However, apart from technical and economic feasibility, other factors have to be considered, such as social and environmental aspects, access to sites linked to security and logistical constraints, and reliability of fuel supplies or the electrical grid (if any). A holistic analysis will be required before hand to see the added value of using solar technology versus other options.

Can solar schemes work well under very high temperatures?

Solar panel efficiency decreases with high temperatures (see more in sections 4.2 and 7.8). This factor is taken into account at the design stage and is therefore not a problem as such. With ambient temperatures of over 55–60°C, resins and epoxies within the solar panel itself could start melting, affecting the functionality of the equipment. However, those temperatures are rarely reached on earth. In summary, solar panels perform sufficiently under high temperatures.

How long is it before solar pumping schemes develop technical problems? And can local communities be trained to carry out repairs?

A well-designed and maintained solar pumping scheme can last a long time (10 years or more) without any major technical issue. However, at some point, these systems can and will develop problems that community technicians won't be able to solve, regardless of the training provided to them. This is why it is important to consider and put in place from the start a model that links users/owners of the system with technical offices or providers with enough technical expertise to solve such issues. It is not recommended to focus on technology alone without taking into account maintenance aspects when it comes to important failures.

CHAPTER 8
Solar-powered water pumping for agriculture

While technology considerations are similar regardless of how water is used, design and operation of solar irrigation schemes come with their own specific challenges. This chapter explains different irrigation techniques and matches them to solar pumping considerations. In addition, it sets out a range of models to help in the financing of solar irrigation schemes, together with a set of the most common risks and problems encountered when using solar pumping for agricultural purposes.

Keywords: irrigated agriculture, solar irrigation, irrigation policy, solar financing, irrigation techniques

8.1 Water for irrigated agriculture

Irrigation is the science of providing water to crops. Irrigation allows farmers to reduce the uncertainty of rainfed farming. With irrigation, farmers can provide the right amount of water needed by the plant depending on the plant growth stage. Reducing uncertainty increases yield and improves livelihoods of farmers (see Figure 8.1).

Agriculture accounts for around 70 per cent of water withdrawals (91 per cent of which is for irrigation) from rivers, lakes, and aquifers (FAO & WWC, 2015). The industrial and municipal sectors account for 20 per cent and 10 per cent respectively (Aquastat, 2019).

Farming activities are usually performed in rural areas where grid connection is non-existent or unreliable. In 2017, an estimated 1.1 billion people – 14 per cent of the global population – did not have access to electricity (IEA, 2017). Many more households experience power outages. Almost 84 per cent of those without electricity access live in rural areas in Sub-Saharan Africa and developing Asia. For this reason, the main energy source for irrigation is fossil fuel (diesel, gas, or petrol). Both liquid fuels and electricity require infrastructure between the producer and consumer.

8.2 The influence of pressure on energy requirements in irrigation

Irrigation techniques can be summarized as six main types:

- *Flood irrigation* delivers water to the crop by flooding the base of the plant. Also known as furrow irrigation, it is achieved by digging a trench between plant rows to deliver water using gravity.

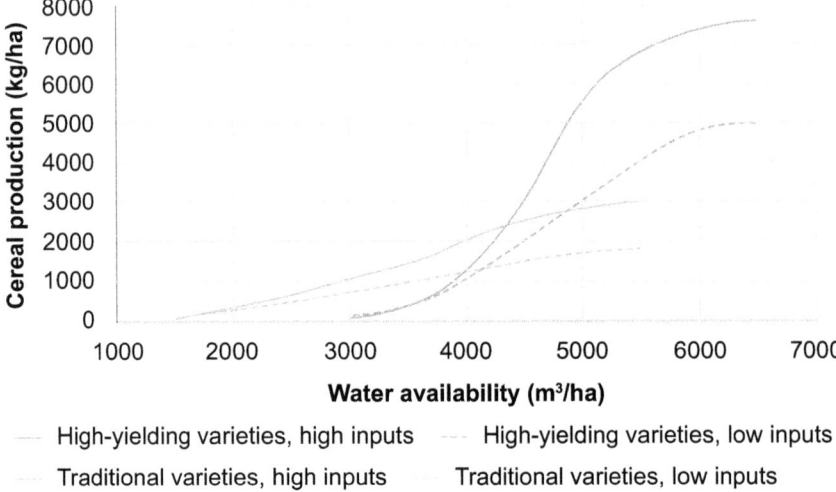

Figure 8.1 Yield response of crops to water availability in China
Source: Smith et al., 2001

- *Drip irrigation* uses pipes installed at the plant base along the plant rows, delivering water drop by drop to the plant.
- *Sprinkler irrigation* sprays water into the air to irrigate the crops using spray heads. It irrigates the entire soil surface.
- *Micro-sprinkler* is a mix of drip and sprinkler irrigation. The micro-sprinkler heads are set up close to the plant base and spray water locally in small water drops. They require the same amount of piping as drip irrigation but use different sprinkler heads. Furthermore, the pipes are elevated off the ground.
- *Pivot irrigation* uses the sprinkler irrigation method but with the spraying head fixed to a machine rotating around a pivot. Circular patterns are created with this technique.
- *Travelling guns* are also derivatives of the sprinkler technology. They operate parallel to the field which needs to be irrigated and at higher pressure than a classic rotating sprinkler.

The estimated pressure requirements for each irrigation technique are:

Irrigation technique	Estimated pressure requirement
Flood	0 bar
Drip	0.5–3 bar
Micro Sprinkler	0.5–2 bar
Pivot	1–3 bar
Sprinkler	2–5 bar
Travelling gun	2–8 bar

Source: FAO & GIZ, 2018b

Pressurized irrigation systems need a tank to operate. The tank height provides the pressure need of the irrigation technique. One bar represents an increase of total dynamic head of 10 m and irrigation tanks are usually set up close to the fields and not higher than 10 m. For this reason, for irrigation techniques requiring pressure of more than 1 bar it could be necessary to add a water booster in the system. Water boosters require electricity to operate. Therefore, solar energy is recommended for low-pressure irrigation techniques, such as flood, drip, micro-sprinkler, and some pivot systems.

8.3 Greenhouse gas emissions from agriculture and climate-change adaptation

The three main sources of greenhouse gas (GHG) emissions in agri-food systems are carbon dioxide from food processing and food waste degradation, methane from ruminants, and nitrous oxide from manure left on pastures. It is estimated that irrigation accounts for 10 per cent of GHG emissions from the agri-food sector (see Figure 8.2). If solar is to replace existing fuel-based energy sources for irrigation this will have a clear impact on the mitigation of GHG emissions in agri-food systems. Once installed, solar irrigation does not produce GHG emissions

The more common solar-powered irrigation systems (SPISs) become, the less greenhouse gases will be released into the atmosphere compared to fossil fuel or grid irrigation. SPISs do not only offer an opportunity for mitigating GHG emissions, but also for adaptation. Irrigating farmers are more resilient against erratic rainfall patterns caused by climate-change effects and as they produce their own electricity they do not depend on an external energy provider (fuel or erratic grid) (FAO, 2017). The main

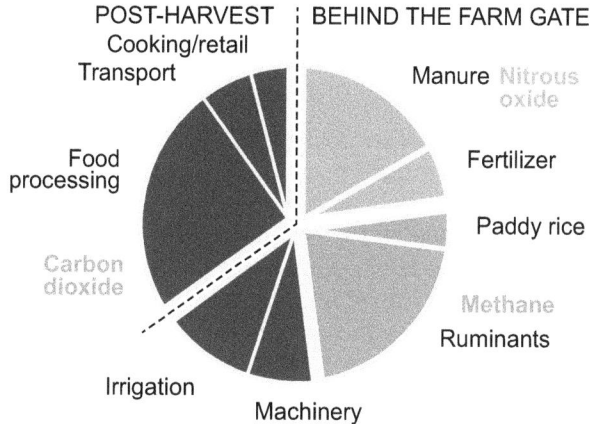

Figure 8.2 Approximate shares of greenhouse gas emissions (in CO_2 equivalent) emitted by the global agri-food sector in 2010
Source: FAO, 2015

benefit of SPISs compared to grid pumping is the reliability of the system, if well maintained. Solar energy farmers become independent energy producers and can schedule their irrigation needs instead of depending on an external energy source.

8.4 Financing solar-powered irrigation systems

One of the main differences between solar pumping for drinking and for agriculture is the nature of the investment. Drinking water is a human right and pumping usually benefits a whole community. Therefore, the group could collectively invest in the pumping system. Individual farmers or cooperatives invest in irrigation systems to support their activity and increase their revenues. Therefore, investment costs are strong drivers for the adoption of SPISs because individual farmers or groups of farmers support them.

The agriculture sector represents 63 per cent of the employment in low-income countries and 55 per cent in Sub-Saharan Africa (World Bank, 2018). Low-income countries are countries with a gross national income per capita of less than US$996 per year (World Bank, 2018). The cost of energy for irrigation (investments and operational costs) is an important parameter for farmers.

The cost of SPIS pumping equipment has been brought down by the plummeting of solar PV prices. Though solar equipment is only one of the elements of an SPIS, it represents one of the most important drivers for adoption by farmers. Investment costs for SPISs remain higher than fuel-based pumping (FAO, 2018). Since SPISs do not require fuel, operational costs are limited to maintenance and replacement at the end of life of the different components. In the fuel-based pumping model, the fuel cost will increase the operational costs over the short to medium term (see more on cost in Chapter 9). These operational costs make fossil fuel–based pumping more expensive than SPISs.

Payback periods for SPISs differ around the world and according to the different subsidy schemes in place (FAO, 2018). Some country examples are:

- *Senegal.* Payback periods for solar pumps in the belt between Dakar and Saint Louis are given as 2–2.5 years (three cropping seasons, 1 ha irrigation with a 15 per cent interest rate on equipment) (Hagenah, 2017).
- *Chile.* The payback period for solar pumps for small farmers (2–4 ha) in northern Chile is around 4 years (three cropping seasons of tomatoes, paprika, and green beans) (R. Schmidt, Africa Solar, pers. comm., 2017).
- *Kenya.* Suppliers of SPISs in Kenya claim a payback period of between 1 and 2 years, sometimes even less (at non-subsidized prices) for SPISs irrigating 1–2 ha of fruit and vegetables (S. Ibrahim, SunCulture, Kenya, pers. comm., 2017).
- *India.* The payback period depends on the state subsidy, ranging from 0 years in Bihar with subsidy for 8 ha of rice, wheat, maize, and lentils

to 19 years without the subsidy. The payback period in West Bengal, where there is no subsidy, is 8 years for 7 ha of rice and vegetables (Mukherji et al., 2017a).
- *Pakistan.* The payback period is 6 years for 32 ha of rice, wheat, and cotton. There is no subsidy on the cost of equipment.
- *Nepal.* A payback period of 2.5 years is reported with subsidy and 8 years without subsidy for 5 ha of rice, wheat, and vegetables. The financing mechanism entails a subsidy (up to 70 per cent of the total cost for a female farmer compared to 60 per cent for a male farmer) and a loan with 15 per cent interest rate (Mukherji et al., 2017b).

SPISs are economical thanks to the short payback period. The systems are dependable because farmers do not rely on the grid or on fuel delivery. However, a number of risks and challenges persist.

8.5 Financing instruments to develop solar irrigation

Farming revenues depend on seasonality and SPISs may require a high initial capital investment, which implies:

- long repayment periods (5–10 years);
- the need for high profit margins on agricultural products or associated services;
- the need for a grace period at the beginning of the repayment plan.

Irrigation brings higher revenues to farmers but the investment and collateral needed to invest in SPISs may be too high for small-scale farmers.

For this reason, farmers need to perform an economic analysis of their farm and to run scenarios without irrigation, with fossil or grid irrigation, and with an SPIS. Based on actual investment costs and potential revenues with irrigation, farmers can define the investment needed as well as the type of financing instrument to use.

Different financing instruments for solar irrigation, discussed in Annex I, are:

- commercial bank loans;
- rural/development bank loans;
- microfinance
- leasing or repurchase agreements;
- financing through agrarian cooperatives;
- informal saving groups;
- pay-per-use business models;
- SPIS subsidy mechanisms;
- matching grants mechanisms.

A blend of different instruments (a loan and a grant, for instance) can also support farmers in designing, acquiring, operating, and maintaining their SPIS.

8.6 The risks and challenges of solar irrigation

One of the main challenges for small-scale farmers are the capital investment costs of SPISs, as farmers typically lack the funds to invest in the system and the collaterals requested by traditional banking. In some cases, a company may be granted a subsidy with a loan component on behalf of a group of farmers and sell water to them using a fee-for-service or a pay-as-you-go model.

The second challenge is related to nature of solar power. Where most electric and fuel pumps deliver a set quantity of water (in m^3/h) from a predefined depth, most solar pumps deliver a set quantity of water according to solar irradiation (unless there is a second energy source in a hybrid configuration). During the day, solar pumps will deliver more water at noon when solar irradiation is at its maximum, whereas it is not recommended to irrigate at midday since evapotranspiration is also at its maximum (FAO, 1998). It is recommended to irrigate when evapotranspiration is lowest, in the morning and evening, at which time the solar pump output is lowest. For farmers, irrigation experts, and extension staff, it is a complete change of paradigm where they have to think in quantity per day instead of hour. This can have an impact on investment costs because farmers may need to install a storage system to be able to irrigate at the right time for their crops.

Another challenge is related to the risk of overpumping since operating costs of SPISs are minimal. State control may play a big role here, in the form of water withdrawal permits or in subsidizing SPISs with water-efficiency equipment. For instance, in Tunisia, subsidies for SPISs are granted only if the farmer is using drip irrigation (FAO & GIZ, 2019). State control also depends on the land and water tenure regime. In cases where landowners are also water owners, the state cannot exert control over groundwater pumping. The absence of state control, low operating costs, and the importance of agricultural water withdrawal may lead to an increase in water consumption and a decrease in sustainable water resources.

An additional challenge of solar irrigation is to operate and maintain the solar components of irrigation systems. These challenges are not different from the ones observed in other chapters of this book. SPISs are usually operated and maintained by irrigation experts, whose energy knowledge may be basic.

As part of the global initiative Powering Agriculture, GIZ together with FAO and Energypedia developed the SPIS toolbox (see FAO & GIZ, 2018b.) in English, French, and Spanish, in order to help find solutions to technical, economic, and sustainability challenges.

> Training farmers, extension services, and irrigation boards has the potential to reduce the misconception of 'more water – more yield' (FAO & ICIMOD, 2019).

To support the use of the toolbox, FAO and GIZ run capacity-building programmes. GIZ also maintains a list of trainers and has developed the community of trainers on solar irrigation. The objective of these activities is to reinforce local capacity on all aspects of SPISs: environmental, technical, and financial, and to cover sustainability aspects of the water-energy-food nexus.

Finally, national and regional dialogues between the different public and private actors or the water-energy-food nexus are necessary.

8.7 Recommendations for solar irrigation challenges

Globally, agriculture is the largest use of water (FAO & WWC, 2015; Aquastat, 2019). By reducing the energy costs of pumping, farmers may not link a price to water and could overpump. This can have important consequences on water resources. For this reason, a number of policy recommendations for solar irrigation can be made.

Solar-powered irrigation has many facets and a wide variety of stakeholders in the areas of agriculture, energy, and groundwater management. There is a lack of understanding and coordination between the different national ministries in charge of the water-energy-food nexus (FAO, 2014), which could lead to competing and conflicting decisions and to suboptimal solutions. To address this issue, government should establish strategies for sustainable energy and water use, and monitor and control water use (through allocations and quotas) in line with the subsidy mechanism, particularly in water-scarce areas. It is also recommended that policymakers develop a map of operating pumps to identify both the farmers that exploit common water resources and the agricultural wells in operation.

8.7.1 Existing water laws

Solar irrigation has become a reliable and affordable technology. A certain number of agricultural water policies have preceded the technology and all stakeholders should ensure compliance with existing water laws, policies, and frameworks. For instance, the development of SPISs must respect existing water allocations, water quotas, water abstraction permits, and well licensing. It is necessary to ensure the means of control (staff number and quality) and legal provisions for illegal wells equipped with SPISs and abstraction rates higher than allocation. It is also essential to ensure the payment of irrigation fees.

8.7.2 Subsidy leverage

Authorities have a strong role to play in controlling the sector and can use the subsidy leverage to reinforce their position. Premiums paid by users could support efficient water management and community systems, and subsidies can also be made conditional on:

- a water management plan;
- a national or federal directory of solar irrigation systems, featuring GPS coordinates, clear description of components, flowrate, and power of the solar components;

- water abstraction permits corresponding to the water allocation of the subsidized SPIS and the type of crops irrigated or agricultural products;
- water and energy metering;
- electronic control devices that transmit real-time information on reservoir levels, pump flows, and water drilling;
- training of subsidy recipients on water efficiency and crop water requirements;
- community or cooperative-owned systems based on volume delivery since collective water abstraction tends to be more efficient than individual water abstraction;
- different subsidy mechanisms according to the type of water source. It should be easier for farmers to access subsidies if they withdraw surface water (easier to monitor decreasing availability) and more difficult if they withdraw groundwater (difficult to monitor depletion).

Policymakers could also establish a register of controllers and make the sale of variable speed drive controllers conditional on the existence of an authorized well and licensing.

8.7.3 Licensing suppliers of SPIS components

Governments are in charge of quality control of solar products for irrigation and other uses. However, in some countries, commercial providers of cheap, low-quality solar-irrigation components are entering the market without proper licensing. Policymakers can limit this through different measures; for instance, the Infrastructure Development Company Limited (IDCOL), a government-owned non-bank financial institution of Bangladesh, has developed a list of approved service providers. Another solution is to establish working groups on solar irrigation within solar industry associations to ensure products comply with minimum quality standards.

In addition, regulatory bodies must ensure companies manage SPIS components at their end of life by setting up schemes for collection, repairs, recovery, recycling, and safe disposal of SPIS products present on the market. As an example, India started a pilot in April 2019 to define and implement a regulatory framework to manage PV module waste (Bridge to India, 2019).

8.7.4 Integrating other uses for solar energy

Agriculture requires energy (electricity or fuel) for different uses, be it machinery, equipment, heating, cooling, or lighting spaces, and indirectly through fertilizers and chemicals (Schnepf, 2006). Solar energy produced in excess of requirements for water pumping could have other on-farm uses:

- drying, milling, milking, husking, grinding, threshing, or pressing of crops as a complementary activity when the water pump is not in use (GIZ, 2016);

- cold storage and lighting if the system is equipped with batteries (FAO, 2018b);
- charging batteries for agricultural machinery;
- household uses (mobile charging, lighting, entertainment, water heating, etc.).

Planning alternative electricity uses supports the valuation of produced electricity to reduce the overuse of pumps and increase revenue streams to invest in SPIS capital expenditure. It could also ensure optimal use of the solar equipment and empower farmers to make rational decisions on the use of energy and water at farm level.

8.7.5 Solar irrigation service providers

Another potential solution to reduce the challenge of financing SPISs is to establish solar irrigation service providers (IWMI, 2018). In this scheme, small companies have their own portable panels, pumps, and solar equipment, and sell water to farmers for a fee. This model exists in Bangladesh, Bihar state (India), and south of Nepal (Terai). It is aimed at replacing existing irrigation water providers that use diesel pumps. In this case, the farmer does not have to invest in the technology and the service provider is maximizing the return on their pump and solar equipment, therefore reducing the payback period of their investment. India also has a different model for grid-tied solar pumping systems, where a cooperative is established to manage the SPIS. This cooperative is registered as an independent power producer and can contract a power purchase agreement with the utility based on the negotiated feed-in tariff (FiT). Farmers are able to make informed decisions on either watering their crops or injecting renewable energy to the grid. They base their decisions on potential cash flow from the produced food or from the FiT.

Solar energy has the potential to revolutionize irrigation. As such, raising the awareness of agricultural experts and farmers on the potential and impacts of solar irrigation is urgent, notably on water stress. Strengthening the capacity of agricultural extension services and vocational training centres on SPISs will lead to improved maintenance and contribute to managing the risks of water stress.

CHAPTER 9
Economic analysis: life-cycle cost of different pumping technologies

Economic considerations, beyond capital costs, are essential when considering different water pumping solutions. This chapter explains life-cycle cost analysis step by step as a way to properly compare costs over time of different pumping technologies. This analysis is presented as a decision-making tool for implementers and communities of users to work out the most cost-efficient way over time to supply water. Examples and tools are given and/or referenced throughout the chapter to facilitate understanding. Finally, different business models suited to financing solar pumping schemes are shown.

Keywords: life cycle cost analysis, payback period, financing models for solar pumping, lease to own, discount rates

9.1 The importance of economic considerations

Economic considerations are important when comparing alternative pumping methods. In many cases, hydrological or climatological factors will limit the kind of pumping system that can be used. Where equivalent performance alternatives exist, their evaluation must include both economic and technical analysis.

Energy for water supply is, in a large number of humanitarian contexts, partially or totally dependent on generators and fuel. Pumping of water is a high energy-consumption activity, resulting in high recurrent costs, particularly for fuel supply and maintenance of equipment. Depending on the local setting, the provision of fuel can be extremely costly and energy-intensive.

While the capital costs for solarizing water points are normally higher than those of other technologies, such as diesel pumping systems, it has been established that adoption of solar energy systems translates to higher savings with time.

The Global Solar and Water Initiative (2018) has analysed the life-cycle economics of 160 water schemes in 55 refugee camps and communities, establishing return-on-investment periods (solar vs generator) of 0 to 4 years and cost reductions over time ranging from 40 to 90 per cent.

These financial estimations vary from water scheme to water scheme however. What will the capital costs be for a solar system, how quickly will the investment be compensated by the savings produced, and how much of the costs will be reduced over time?

These are questions that need to be answered before starting any work on the ground in order to know how worthwhile, from the financial point of

134 SOLAR PUMPING FOR WATER SUPPLY

view, it is to invest in solar equipment, as well as to prioritize those schemes that produce higher savings more quickly.

9.2 Life-cycle cost analysis

9.2.1 Main concepts

The most complete approach to economic appraisal when comparing different pumping technologies is the life-cycle cost analysis (LCCA). In this analysis all future costs and benefits are calculated in today's money value. Because the value of money changes over time it would be unrealistic to add up costs incurred in different years. Instead, those costs need to be converted into money value at the same point in time, normally the present time. A life-cycle cost analysis considers the following concepts.

Life-cycle costs. The sum of all costs and benefits associated with the pumping system over its lifetime (or over a selected period of analysis), expressed in present-day money. This is called the *present worth* or the *net present value* of the system.

Payback period. The length of time required for the initial investment to be repaid by the benefits gained.

Total cost saving. The difference in total costs incurred between two different investments at the end of the appraisal period.

Discount factor (also called real interest rate). An index that expresses the change in value of money over time in a certain country for a certain product. This is not the change due to general inflation, but the difference in return between a chosen investment and another that is not chosen (e.g. if a lender is receiving 9 per cent from a loan and the inflation rate is 8 per cent, then the real interest rate = nominal interest rate – real inflation rate = 9 – 8 = 1 per cent).

9.2.2 Present worth

The present worth is a calculation of all costs and benefits incurred in today's money value. For a payment of Cr($) to be made in the future, the present worth (PW) is found by multiplying the payment Cr by a factor Pr:

(formula 1.1) PW = Cr × Pr, with Pr = $1/(1 + d)^N$

with time for the payment (N, in years) and discount rate (d) as the main variables (note: if d = 12 per cent, d = 0.12 in formula 1.1).

In an ideal situation, there would be a real interest rate difference for every product, since the price of solar PV panels, pumps, generators, diesel, and the like will increase or decrease over time in different ways. However, as it is difficult to have this intimate knowledge of the national market and its price evolution for different products it is common to apply a general discount rate to all products.

Therefore, the real interest rate is calculated by subtracting the real inflation rate from the nominal interest rate, both taken for the country where we are considering the investment (for more values of real interest rates per country see World Bank, 2019.)

In the case that no information can be found for a particular country, a good approximation to the real interest rate would be the commercial bank lending interest rate of the country of work.

So overall the total present worth will be then,

(formula 1.2) Total PW $= I + \sum_{n=1}^{N} Cr \times [(1/(1 + d)^N]$

with I = initial or capital costs and Cr = all other costs incurred over time.

Example 9.1: Calculating the present worth

A bicycle is bought for US$125. After two years of use, the front wheel needs to be changed at a cost of $15. In the third year, changing the brakes costs $10. What is the total cost incurred to buy and maintain the bicycle in today's money value if the real interest rate is 12 per cent?

The costs in different years cannot simply be added together, since the value of money changes over time. Instead, all future costs will be brought to today's value so that they can be added together.

Following formula 1.2,

I = $125; d = 0.12; Cr for Year 1 = 0; Cr for Year 2 = $15; Cr for Year 3 = $10

PW = 125 + 0 + 15 × [(1/(1+0.12)^2] + 10 × [(1/(1+0.12)^3]

= 125 + 12 + 7 = $144, in today's value.

This formula can be added to an Excel sheet, making it easy to consider as many costs and benefits as needed for as many years as needed (see Economic analysis Excel tool at GLOSWI, 2018).

As explained earlier, it is also possible to consider different discount rates for different costs. This might be useful when there is a good understanding about price variation over time for different costs incurred. In Example 9.1, if it is known that bicycle wheel prices are increasing while brakes are decreasing, different real interest rates could be used in the formula for each of them in order to get the present worth. When this is not known (as is often the case) or for the sake of simplicity, a discount rate for the country is found and applied to all the costs, as in the example.

9.3 Life-cycle costing for water pumping

9.3.1 Types of cost to be considered

For every pumping system for which a present worth LCCA will be carried out, all related costs will need to be identified first. These can be generally divided into the following four categories:

- initial capital costs (including installation);
- operation and maintenance (minor and major services and fuel);

136 SOLAR PUMPING FOR WATER SUPPLY

- overhaul;
- replacement of equipment during lifetime of the water scheme.

Once the economic appraisal is done for each of the pumping technologies that could be used for a particular water point, these will be compared in order to establish which one offers more cost savings over time.

A rural water supply system will have five main components: water point (drilling, fencing etc.), power source (solar, generator), pump, water tank, and distribution system (i.e. piping system). Since the point of the LCCA is to establish a comparison between technologies, costing elements that are common for all the different pumping options (e.g. distribution system, drilling, fencing, and others) can be excluded from the analysis in order to simplify it, since they will have the same effect for the different options.

Cost can be divided into capital and future or recurrent costs:

Capital cost. These are the costs of buying the different components of the system, transporting them to the site, and installing them.

Future or recurrent costs. These are the operating costs (e.g. diesel), maintenance or repair costs (e.g. replacement of parts, carrying out repairs), and replacement costs (e.g. components that are replaced when they reach the end of their lifetime).

Figure 9.1 describes the steps for an economic appraisal. Steps 1 to 3 have to do with the technical design of the pumping system, as explained in Chapter 5. These steps are necessary in order to determine the different technologies and equipment sizes that could be used for a particular water point.

Once the possible technical designs have been established, the capital cost of each item needs to be determined by asking suppliers, partner organizations or others for quotations. Similarly, O&M, replacement, and other costs can be estimated through past project experience.

The data required for the last three steps (steps 4 to 6) are given in Table 9.1.

Table 9.1 Data required for life-cycle analysis

Economic	Period of analysis (typically all systems are taken to the longest lifespan of any of the components, which is 25 years for solar panels)
	Discount rate (= nominal interest rate – inflation rate)
	Relative inflation rate (typically zero)
Cost of each component	Capital cost[1]
	Annual O&M, overhaul, and replacement costs
	Labour costs
Technical	Lifetime of each component

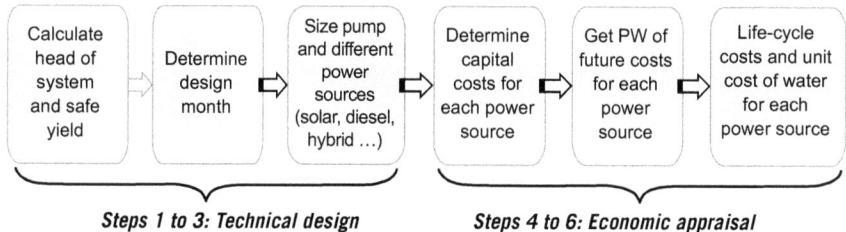

Figure 9.1 Steps for technical design and economic appraisal

9.3.2 Guidance costs and values

In case the manufacturer does not state lifetime information clearly on their datasheets, the following can be used as a guide as to when different system components will likely need to be replaced, based on good-quality products in compliance with manufacturing certifications:

- solar panels: 25 years;
- pump ends and motor: 10 and 7 years respectively;
- inverter / control equipment: 7 years;
- civil structure and frames: 25 years;
- mechanical and electrical fittings: 25 years.

Similarly, for diesel generators, where field information is not available regarding replacement frequency and costs, Table 9.2 provided by a good-quality generator manufacturer (and considered from experience as a conservative estimate) can be used as a reference.

Finally, fuel consumption for a generator will need to be established in order to estimate running costs. Again, the best way to get this data is from previous field experience. Where this is not available, Table 9.3, provided by a good-quality generator manufacturer, can be referred to.

Once the LCCA is performed between diesel, solar, and any other pumping option being considered, a choice will have to be made. Of course, the least cost solution may not be the final choice once other factors are taken into account; reliability or ease of maintaining equipment, for example, could be the most important considerations and users may be prepared to pay the extra cost. However, as said earlier, a cost comparison is a necessary step before making a choice of technology.

Table 9.2 Estimated cost for maintenance of a good-quality diesel generator

Maintenance and replacement	Frequency of change (in working hours of generator)	Price (US$)
Minor service	250	20
Major service	1,000	180
Overhaul	10,000	30 per cent of new generator
Replacement	35,000	Buy new generator

Table 9.3 Estimated generator fuel consumption at different loads

Generator fuel consumption chart – average for good quality engines					
kVA	kW	Load 25%	Load 50%	Load 75%	Load 100%
25	20	2.3	3.4	4.9	6.0
38	30	4.2	6.8	9.1	11.0
50	40	6.0	8.7	12.1	15.1
75	60	6.8	11.0	14.4	18.1
94	75	9.1	12.9	17.4	23.1
125	100	9.8	15.5	21.9	28.0
156	125	11.7	18.9	26.8	34.4
169	135	12.5	20.4	28.7	37.0
188	150	13.6	22.3	31.8	41.2
219	175	15.5	25.7	36.7	48.0
250	200	17.1	29.1	41.6	54.4

Note: A quick rule of thumb for estimating fuel consumption is 0.3–0.5 litres per kWh consumed at pump level.
Source: Genset

9.4 Comparing LCCA of solar and generator systems

9.4.1 Stand-alone generator vs stand-alone solar system: a worked example

A new borehole is drilled and it is calculated that a diesel generator of 15 kVA with a pump of 5.5 kW could supply the daily water needed. However, a solar system of 11 kWp of solar panels working with the same pump could supply the same amount of water.

A present worth life-cycle cost analysis is performed in order to determine the most cost-efficient option. In order to simplify the analysis, costs that are common to both systems are not considered (e.g. the cost of replacing the water pump, cost of fencing the water point, cost of water point guards).

The cost of buying and installing the generator-powered pumping system is $8,450. The generator should work seven hours a day in order to provide the water needed; the costs of operating and maintaining the generator are taken from Table 9.2. The cost of 1 litre of diesel is $1.10 while the discount rate for the country where the system is to be installed is 12 per cent (d = 0.12).

All the associated costs are added up and multiplied by the corresponding Pr factor for each year. The Pr factor expresses not only the increase of price due to inflation, but measures the weight in our budget for paying costs now and in the future. A high discount rate (d) would mean that costs in the first years will weigh more in the budget than future costs, and vice versa. The LCCA for the scheme with a diesel generator is given in Table 9.4.

The capital cost of the solar pumping system is given in Table 9.5.

Repeating the analysis for the solar option, where no fuel is used, the only replacement to be done is that of the inverter/control box, every seven years

ECONOMIC ANALYSIS: LIFE-CYCLE COST OF DIFFERENT PUMPING

Table 9.4 LCCA for the given water scheme with a diesel generator

Year (N)	Capital cost of generator system ($)	Generator working time (h/year)	Minor service $20 every 250 h	Major service $180 every 1,000 h	Fuel consumption (litres/h)	Cost of fuel ($)	Overhaul 30% generator cost every 10,000 h ($)	Replacement New genset every 35,000 h ($)	Pr = $1/(1+d)^N$	Total costs in present worth ($)
0	8,450	2,555	204	460	3.5	9,837			1.000	18,951
1		2,555	204	460		9,837	0	0	0.893	9,376
2		2,555	204	460		9,837	0	0	0.797	8,371
3		2,555	204	460		9,837	1,350	0	0.712	8,435
4		2,555	204	460		9,837	0	0	0.636	6,674
5		2,555	204	460		9,837	0	0	0.567	5,959
6		2,555	204	460		9,837	0	0	0.507	5,320
7		2,555	204	460		9,837	1,350	0	0.452	5,361
8		2,555	204	460		9,837	0	0	0.404	4,241
9		2,555	204	460		9,837	0	0	0.361	3,787
10		2,555	204	460		9,837	0	0	0.322	3,381
11		2,555	204	460		9,837	1,350	0	0.287	3,407
12		2,555	204	460		9,837	0	0	0.257	2,695
13		2,555	204	460		9,837	0	4,500	0.229	3,438
14		2,555	204	460		9,837	0	0	0.205	2,149
15		2,555	204	460		9,837	0	0	0.183	1,919
16		2,555	204	460		9,837	0	0	0.163	1,713
17		2,555	204	460		9,837	1,350	0	0.146	1,726
18		2,555	204	460		9,837	0	0	0.130	1,366
19		2,555	204	460		9,837	0	0	0.116	1,219
20		2,555	204	460		9,837	0	0	0.104	1,089
21		2,555	204	460		9,837	1,350	0	0.093	1,097
22		2,555	204	460		9,837	0	0	0.083	868
23		2,555	204	460		9,837	0	0	0.074	775
24		2,555	204	460		9,837	0	0	0.066	692
								Total cost in present worth		$104,007

as standard lifetime. An additional annual cost of $1,500 is included to cater for the cost of preventive maintenance, regular cleaning of the solar panels, and any small repairs that may be needed over time. The LCCA for a solar scheme is given in Table 9.6.

Additional costs that need to be factored in can be added in a new column with the analysis proceeding in the same way. Remember, this is a comparison between technologies and therefore costs that are similar to both solutions (e.g. salary of the water point operator) can be disregarded in order to simplify the analysis.

140 SOLAR PUMPING FOR WATER SUPPLY

Table 9.5 Capital cost of the main components of the given solar PV pumping system

Component	Unit	Quantity	Unit price ($)	Total price ($)
Pump	W	5,500	0.58	3,200
Inverter	W	5,500	0.33	1,800
Solar modules	W	11,000	0.80	8,800
DC accessories	W	11,000	0.12	1,320
Cables and low-level sensors	m	60	7.00	420
PVC pipes and wellhead cover	m	90	5.00	450
Support structure	W	11,000	0.35	3,850
Subtotal				19,840
Add 10% for installation				1,984
Total				**21,824**

Table 9.6 LCCA for the given water scheme with a solar PV pumping system

Year (N)	Capital cost ($)	Preventive and minor service and cleaning ($)	Major service NA	Fuel consumption (litres/h)	Cost of fuel ($)	Overhaul ($)	Replacement cost Change invertor every 7 years ($)	Pr = 1/(1+d)^N	Total costs in present worth ($)
0	21,824	1,500	0	0	0			1.000	23,324
1		1,500	0		0			0.893	1,339
2		1,500	0		0			0.797	1,196
3		1,500	0		0			0.712	1,068
4		1,500	0		0			0.636	953
5		1,500	0		0			0.567	851
6		1,500	0		0		1,800	0.507	1,672
7		1,500	0		0			0.452	679
8		1,500	0		0			0.404	606
9		1,500	0		0			0.361	541
10		1,500	0		0			0.322	483
11		1,500	0		0			0.287	431
12		1,500	0		0			0.257	385
13		1,500	0		0		1,800	0.229	756
14		1,500	0		0			0.205	307
15		1,500	0		0			0.183	274
16		1,500	0		0			0.163	245
17		1,500	0		0			0.146	218

ECONOMIC ANALYSIS: LIFE-CYCLE COST OF DIFFERENT PUMPING

Year (N)	Capital cost ($)	Preventive and minor service and cleaning ($)	Major service NA	Fuel consumption (litres/h)	Cost of fuel ($)	Overhaul ($)	Replacement cost Change invertor every 7 years ($)	Pr = 1/(1+d)^N	Total costs in present worth ($)
18		1,500	0		0			0.130	195
19		1,500	0		0			0.116	174
20		1,500	0		0			0.104	156
21		1,500	0		0		1,800	0.093	305
22		1,500	0		0			0.083	124
23		1,500	0		0			0.074	111
24		1,500	0		0			0.066	99
							Total cost in present worth		36,492

Figure 9.2 plots the cumulative costs for both systems. As shown, there is a considerable cost reduction over the lifetime of the equipment (65 per cent) when the generator is replaced by a solar solution at that particular water point. It is also common to look at shorter periods of time when considering

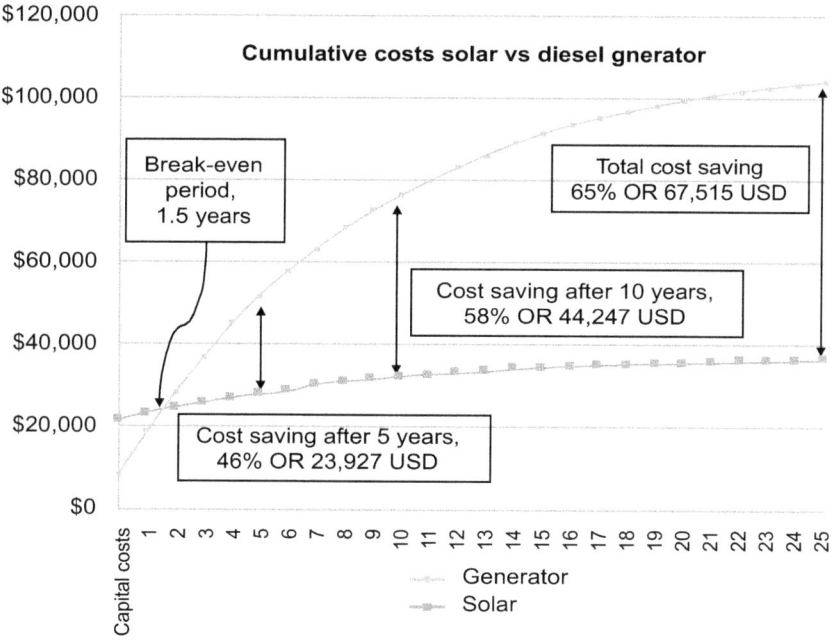

Figure 9.2 Comparison over time of costs for solar vs generator

solar investment. Figure 9.2 shows that savings incurred grow quickly and in year 5 they are already 46 per cent.

Since there are a number of variable factors (e.g. working hours of generator, size of the systems, daily water requirements) and there may be a wide difference in terms of potential cost savings and break-even periods among the different water schemes in the same area of work, it is always useful to perform a similar LCCA for all water points to prioritize intervention in contexts where funding is limited. Table 9.7, compiled during a visit by the authors to South Sudan illustrates these variations, with a wide range in cost savings and break-even periods.

In addition, it is not uncommon to find break-even points of solar compared to diesel generators of one year or less, making the investment worthwhile even for donors with narrow funding windows. Another point to bear in mind and not considered in the analysis is that, in many contexts, diesel is considered a commodity and therefore theft is widespread during transport and storage, a problem that is minimized or eliminated through the adoption of stand-alone solar solutions.

> Although not quantified in economic analysis, solarizing boreholes minimizes or eliminates the cost linked to diesel theft, a widespread problem in many humanitarian contexts.

A final consideration is that the more remote the water point is, the more expensive it will be to transport fuel, increasing the cost of running a generator, while solar costs would remain more or less the same over time.

9.4.2 Stand-alone generator vs hybrid (solar and generator) systems

As discussed in other chapters, sometimes a stand-alone solar system cannot provide the total daily water required. It may then be necessary to consider either a stand-alone generator or a hybrid system that would be solar powered during the solar day and powered by generator for some night hours.

In order to find which of these two options is more cost-effective, an LCCA can be performed, calculating on one side the cost over time of pumping all the water required with a stand-alone generator and, on the other, the cost of pumping during the solar day with a solar system plus the cost of using a generator for some additional hours to provide all the water required.

If for any reason a diesel generator cannot be considered (e.g. difficult to transport diesel to the site because of security reasons, or bad roads) and a stand-alone solar system cannot provide all the water needed, drilling a second borehole equipped with solar panels could be considered.

In Figure 9.3 the three options are considered for a refugee camp in Tanzania in 2016. Normally, a diesel-free option will be more cost-effective in the medium and long term.

Table 9.7 Cost comparison between existing generator stand-alone systems in South Sudan and equivalent solar or hybrid systems

	Site details			Water output			Economic/life cycle analysis				Hybrid/Solar - Diesel Comparison	
							Generator stand alone		Solar stand alone or Hybrid			
No.	Camp	BH ID	Managing agency	Average daily output of prposed solar (m^3/day)	Daily output of proposed generator to work with solar, m^3/d	Daily output of current generator (m^3/day)	Initial cost (USD)	Cost over life cycle (USD)	Initial cost (USD)	Cost over life cycle (USD)	Reduction of expenses hybrid/solar vs Genset	Break-even point
1	Bentiu	Sector 1 Block 7	Mercy Corps	71.1	56.1	127.2	$10,013	$357,425	$12,858	$238,688	−33%	0.7 years
2	Bentiu	Sector 2 Block 9	IOM	120.0	0.0	120.0	$12,468	$371,939	$30,767	$194,943	−48%	1.6 years
3	Bentiu	Sector 3 Block 1	Concern Worldwide	151.1	146.0	297.1	$18,895	$660,400	$22,696	$467,750	−29%	1.2 years
4	Ajuong Thok	Market	Samaritan Purse	72.0	0.0	72.0	$17,200	$272,981	$21,500	$34,515	−87%	0.2 years

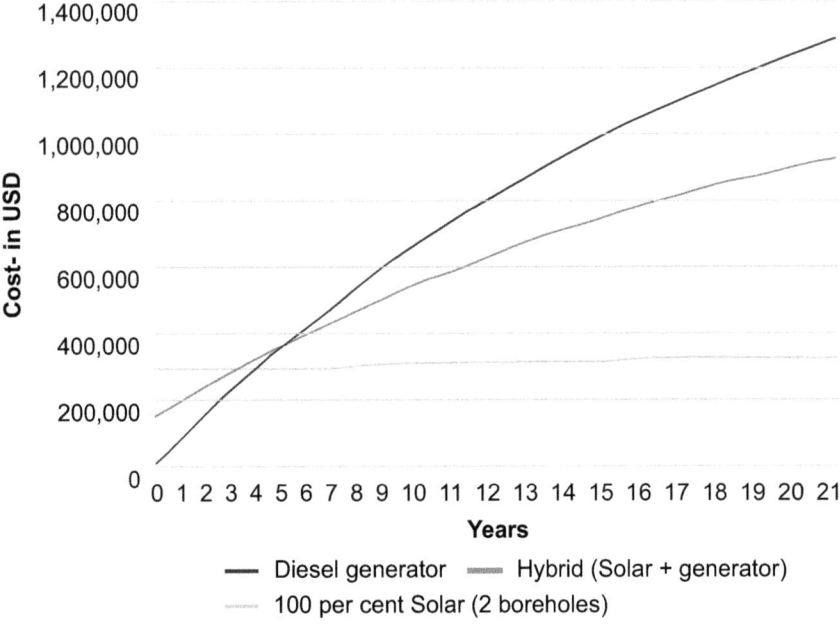

Figure 9.3 LCCA of stand-alone generator vs hybrid vs solar at Nyarugusu, Tanzania

9.5 Cost of ownership

Different business models can be adopted to finance solar-powered water projects making them affordable to more people and therefore accessible (see Figure 9.4).

9.5.1 Direct cash or financed purchase

The organization in charge hires a contractor to design, procure, and install the system. Costs are paid at once in cash; a financing partner may be

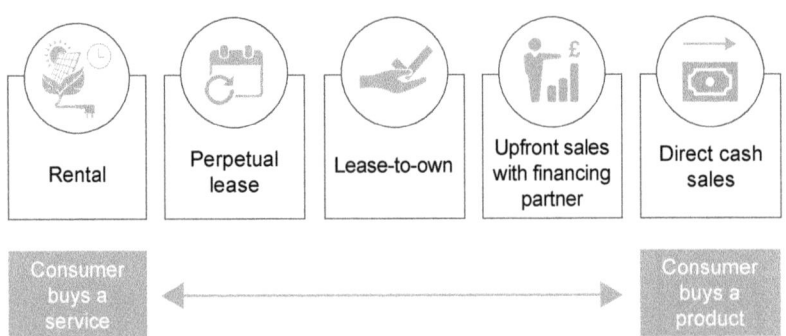

Figure 9.4 Business models for financing solar-powered water systems

involved if cash is only partially available. Once installed, the ownership and management remains with the organization in charge or is transferred to the users or local government entity. Any of those stakeholders can take in hand the operation and maintenance of the system or outsource it to a third party for a fee.

9.5.2 Upfront sales with financing partner

The organization in charge partners with a financial entity that provides funds to interested partners to build a solar scheme; these funds can then be reimbursed in a pre-negotiated way so that the burden of upfront costs are eased for all parties involved.

9.5.3 Lease to own

The organization in charge leases (or rents out) the solar pumping system from a solar service company and pays a monthly fee. The solar service company will finance design, installation, and operation and maintenance of the system. The ownership of the system is transferred to the organization in charge when the lease ends, which can then outsource the operations and maintenance of the system for a fee. Typically lease agreements can last from two to three years up to ten years.

9.5.4 Perpetual lease or power (or water) purchase agreement

The organization in charge buys solar energy (or water) and pays a monthly bill. A power purchase agreement (PPA) will cover only the energy generated (or water pumped) by the solar system. The solar service company will finance design, installation, and operation and maintenance of the system. At the end of the PPA, the organization can renew or cancel the agreement with the solar service company remaining as the owner of the equipment in any case. PPAs can last from two to three years up to ten years.

9.5.5 Rental

The organization in charge rents the equipment from a solar service company for a fixed monthly amount. The solar service company will finance design, installation, and possibly operation and maintenance of the system. At the end of the rental period agreement the organization can renew or cancel the agreement with the solar service company remaining the owner of the equipment in any case.

CHAPTER 10
Calls for proposal and bidding

Quality considerations in the selection of solar components are critical to achieve the water outputs expected. Manufacturing certification criteria are the best way to ensure quality components are selected. This chapter also describes other aspects of solar pumping purchasing and bidding, including warranty, deliverables, and tender evaluation. A template for solar pumping bidding is provided. In addition, the chapter explains the procedures and tools available online to ensure manufacturing certification conformity.

Keywords: solar quality performance, IEC/EN 61215 and 61730, solar bidding, linear performance warranty, solar pumping toolkit, certification database, supplier selection, solar product authenticity

10.1 Selection criteria for solar pumping products and services

Quality is a key criterion in equipment selection and can be defined as the ability of a product or service to consistently fulfil defined requirements over a defined period. Quality is an integral part of a product or service and should be a key consideration in solar pumping equipment selection.

Selection of products of dubious quality (especially related to solar modules) may result in declining performance of the pumping system that may erroneously cast doubt over the overall suitability and sustainability of the project. On the other hand, carefully selected quality products will ensure value for money while contributing to the sustainability of the project.

Some dimensions of product quality that should be considered to establish the desired characteristics of a solar water pumping system are described in Box 10.1.

10.2 Desired features of key components

Table 10.1 summarizes the key minimum or desirable characteristics of solar pumping equipment.

All equipment should be inspected in the supplier's warehouse before delivery onsite to identify and eliminate poor-quality and damaged goods. They should also be inspected and verified on delivery to site. The authenticity of the product should also be checked by looking at equipment serial numbers and certifications and sending these to the original manufacturer for verification. Solar modules can be verified on certificate databases, discussed in section 10.6.

http://dx.doi.org/10.3362/9781780447810.010

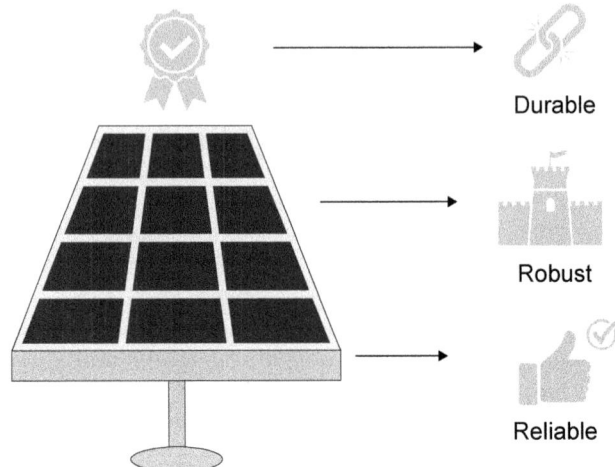

Figure 10.1 Quality of solar panels matters

Box 10.1 Dimensions of product quality

Performance The selected product must meet the specified water demand and head during the design month, which can be the month with the least irradiation, the driest month etc. The design provided by the bidders should show the monthly pumped outputs in line with the duty requirements.

Efficiency Efficiency defines the amount of water that will be pumped with the available power and it impacts the cost of ownership. The payback period for the solar system with the expected cost savings over the life cycle of the system will be a differentiating factor of systems with similar performance. With an efficient system you either spend less for the same amount of water or get more water from the same power source. (see more at Lorentz, 2020a).

Reliability The desired product should not fail within a specific period. Evidence that the product has worked in the past without failure from field examples will be a good indicator of reliability. References provided by the supplier and actual site visits will confirm this. Manufacturer warranty is also a display of manufacturer confidence in the reliability of the product.

Conformity The product should meet the specified standards of quality and safety such as IEC/EN certifications, UL/MET listing, or other local and internationally recognized standards. The supplier should provide verifiable certificates of conformity.

Durability Different components will be able able to exist for a long time without significant deterioration. Solar modules will be typically warrantied for 25 years. Quality pumps and controllers have 7 to 10 year life span.

Robustness	The system should be designed for use in remote locations without failure. The particular environment the system will be installed in should be accommodated for at the design stage, e.g. an enclosure rating for high temperature conditions, or coastal salt mist conditions. Attention should be paid to the specifications given by the supplier.
Serviceability	All components should be subject to minimal servicing and without expensive parts. Parts should only need to be replaced infrequently to reduce replacement costs. The expected time and skill level required for replacement of parts should be such that downtime is reduced.
Aesthetics	Even though it has no bearing on performance, the installed system should be pleasing to the eye as this can enhance community acceptance.

Table 10.1 Minimum recommended characteristics of solar components

Criteria	Equipment characteristics		
	Solar module	Pump and motor	Controller
Performance	PV modules to meet the power demand of the pump	Positive displacement or centrifugal pumps Variable-speed induction motor that can overcome harmonics (motor insulation PE2/PA) Duty point of the pumping system should meet the requirement	Maximum power point tracking and current boosting function Rated to meet the current, voltage, and power demand of the pump
Efficiency	>14 per cent	Motor: >80 per cent	>95 per cent
Reliability	Prior successful off-grid use	Used previously in other sites without failure	Used previously in other sites without failure
Conformity	IEC/EN 61215 & 61730 for crystalline modules	IEC/EN 809, 60034-1, 62253, 61702 ISO certifications	IEC/EN 61800-1, 61800-3, 60204-1, 61683, 62109, 62093
	IEC 61646 for thin-film PV modules		ISO certifications
	UL 1703 listed		
	IEC/EN 61701 (for coastal areas)		
Durability	25 year performance guarantee 20 year product warranty	7 years for motor 10 years for pump	7 years

(Continued)

Table 10.1 Continued

Robustness	Quality encapsulation	Constructed of non-corrodible material, such as stainless steel (AISI 304 or higher)	Environmental protection of IP54 or higher
	Do not break easily		Closely matched to actual conditions (e.g. ambient temperature, humidity, air salinity, altitude)
	Sturdy frame	Permanently lubricated	
		Closely matched to groundwater temperature and water quality	
Serviceability	Terminal/junction box on the module should have a provision for 'opening' for replacing the cable, if required	Modular design i.e. detachable pump and motor	Preferably keep electronics above ground (not submerged)
		Maintenance free where possible (e.g. brushless motors)	Low replacement cost for spare parts
			Low frequency of repair

10.3 Supplier selection

Besides products, selecting a good supplier is critical to having a sustainable solar-powered water system. The right supplier will not only provide the right product but also provide sound installation and workmanship as well as after-sales support, which will contribute to system longevity. Some criteria that may be applied in selecting a good supplier are presented in Box 10.2.

Box 10.2 Criteria for supplier selection

Product quality	Operate a quality management system that is ISO 9000 or equivalent and have recognized third-party verification. Have UL/MET listed products for supply
Delivery	Reliable supplier that can deliver within the required time frame
Experience	Relevant experience of designing, installing, and maintaining solar pumping solutions of a similar size, scope, and application
Reputation	Good international standing in the industry. Provide references from previous customers and peers
Capacity	Staff, tools, and equipment to satisfactorily execute the project
Technical capability	Certified trained staff who can successfully install the system and service the project
Warranty and after-sales service	Policies that support post-installation replacements and repairs. Must have access to spare parts supply with backing from the equipment manufacturer
Training	Certified trained staff who can provide training to the organization, operators, and users

Potential suppliers should be visited to gauge the quality and quantity of the products they stock, to discuss their delivery lead times for non-stock items, and to gauge their overall capability and capacity to deliver the project in a timely manner. It is also advisable to get in touch with their referees and, where possible, visit project sites that they have implemented previously as all this will reveal unscrupulous suppliers.

10.4 Bidding process

Most SPWS projects will require multiple bidders to bid before a contract is awarded. Usually, potential bidders will be asked to submit a proposal according to the requirements set out on a bidding document. The bidding document will state the statutory legal requirements, financial requirements, and technical requirements. This section focuses on bidding for the technical side.

There are two approaches to carrying out the technical bidding process.

1. One involves the buying organization carrying out the technical design of the project and issuing bidders with a list of preferred equipment to be supplied, including brands/models (or equivalents), specific equipment sizes, power ratings, and quantities. This approach is simple and allows supplier selection to be done quickly, though it should be used only where the organization's technical expertise is good enough to do the design and to properly evaluate bids, or where an experienced consultant or private contractor can be engaged to carry out the design. This approach is especially useful when procurement needs to be done speedily, such as in emergency response situations where water supply systems need to be installed within the shortest time. The emergency kits mentioned in section 7.4 can be used for such emergency contexts (a full list of equipment required for each kit is provided at GLOSWI, 2018e).
2. The second approach involves the organization providing information about the project, location, water demand, and other requirements, and asking bidders to submit proposals (including solar scheme designs) to meet those requirements. This approach may result in a lengthy, tedious, and complex process of filtering through the different proposals to identify the ones that match the requirements. The clearer the requirements are explained and the more accurate the data provided is, the easier and fairer the bidder selection process will be. The bidder that provides a proposal which most closely matches the requirements is selected. This approach also requires the organization to have some reasonable level of technical expertise to be able to evaluate the different proposals, but less so than in the first case.

Internal procurement guidelines will also dictate which approach is acceptable. For example, some organizations require that equipment brands are not mentioned in the request for proposal which can hinder the first approach, although care can be taken to comply with such rules while still providing a full list of the required equipment.

The bidding template in section 10.5 is a guideline on using the second approach, where bidders are asked to provide proposals in response to the provided requirements.

> **Box 10.3 A reference guide for the SPWS procurement process**
>
> The bidding process can be complex and difficult, which, if not well managed, can result in project delays and unsatisfactory delivery of the project. The World Bank document 'Photovoltaics for Community Service Facilities' highlights common issues encountered in the procurement process and how to address them (World Bank, 2010). This document should be referred to during the procurement process for SPWS projects.

10.5 Bidding template: technical terms of reference

The terms of reference for the technical requirements of a bid are outlined here.

10.5.1 General information

Location. Location is a critical requirement when an SWPS is being designed as it is the basis for having the correct solar resources data, such as irradiation, peak sun hours, and temperature. The location should be provided in the form of GPS coordinates.

Background information. This section should give the bidders background information on the project. It lays out the context of the project so that bidders can conceptualize the needs, the current situation, the requirements of the proposed project, and the targeted achievements.

Objectives. The objectives of the project should be clearly laid down here, for example, 'the project aims to provide a power system for water pumping in order to maximize the reduction of diesel fuel demand by pumping as much water as possible through a solar pumping solution'.

System planning and design. Information on system planning and design should be provided, such as what design tools should be used, the design factors to be applied (e.g. losses factored into the design), whether a cost–benefit analysis is required with the proposal, and whether datasheets are required to be submitted with the proposal.

10.5.2 Project information

Scope of work. The scope of work description entails delivery schedule, installation, and commissioning guidelines. It should state in detail what is expected of the supplier, from what the extent of the work will be to what will constitute a completed installation. Details such as workmanship, training, documentation, and lifetime requirements need to be clearly explained.

Water requirements. Water demand requirements will be specified, stating the design approach to be used, such as the worst month method discussed in section 5.3.3.

Water source. The water source and other environmental factors should be well detailed such that the bidder has the complete information required for a sound system design. This includes water-source characteristics and site information, such as distances and tank size.

Other design parameters. Hours of operation and water collection patterns, among other things, are useful for the design process.

System configuration. The bid should provide the configuration design of the system, such as stand-alone solar or hybrid solar–diesel.

A schematic layout of the site can also be provided, such as the one shown in Figure 10.2.

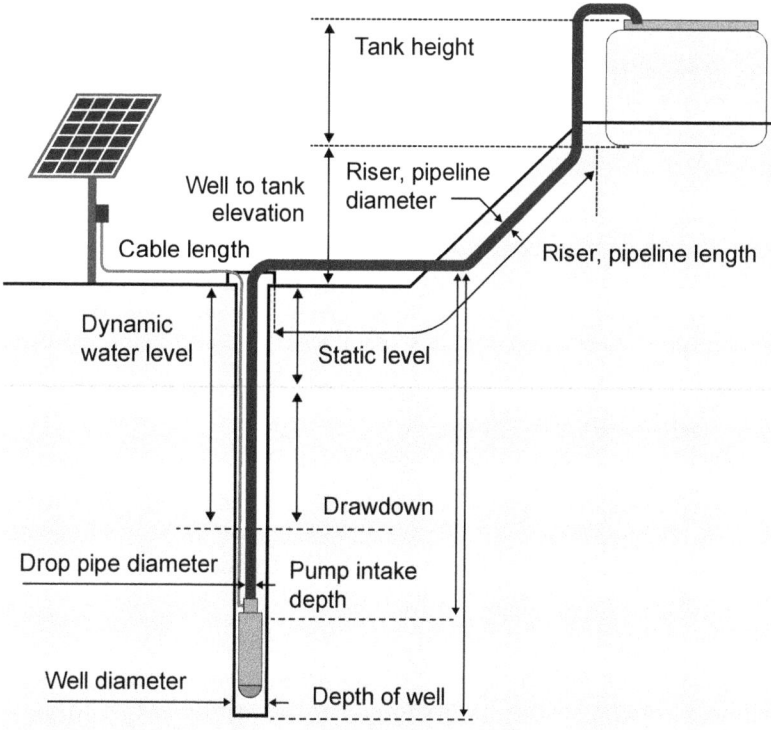

Figure 10.2 Example of a solar-powered water system layout provided to bidders

As a minimum, the information about the project listed in Table 5.1 must be provided for a comprehensive technical design to be carried out.

10.5.3 Equipment specifications

Servicing requirements. The bidders should be asked to state clearly the servicing requirements of the equipment proposed, i.e. the service intervals, cost of parts, time required for replacement, and skill level required for such service.

Spare parts requirements. The bidders should be asked to list the spare parts required, frequency of replacement, and how readily available these parts are.

User friendliness. The equipment must be user friendly for ease of operation and fault finding.

Solar PV modules. The selection of PV modules needs to be given critical attention due to the market proliferation of modules of poorer quality in recent years. Selecting a quality module can only be achieved by meeting the criteria discussed in sections 10.2 and 10.6. The quality requirements for the modules should be provided to the bidders, who are then expected to clearly and in a detailed manner provide a statement of conformity to the quality criteria. Module power rating can also be included. Protection provisions against theft should also be requested, such as permanent marking on the back of the module.

System protection. Include requirements for protection against dry running, surge protection, lightning protection, cable protection, and so on in the equipment specifications.

Control equipment. The equipment controller is the heart of any solar pumping system and should be carefully selected to suit the application. This can be achieved by clearly defining the controller requirements, such as power conversion from DC to AC, efficiency, display interface, quality standards, control inputs, protection features, data logging for remote monitoring, servicing requirements, and environmental protection.

Ancillary control. This is required for a professional installation, such as disconnection switches, combiner boxes, and junction boxes.

Electric pump and motor. Specifications for the required pump will be provided here, such as pump type (AC vs DC, submersible vs surface), pump design (helical vs centrifugal), motor features, pump construction, motor construction, variable-frequency requirements (e.g. PE2/PA motors), efficiency, and quality standards, as stated in section 10.2.

Module support structure. Bidders should be asked to either provide a support structure according to the specifications set out in the bid or to design a support structure that is suitable for the conditions set out in the bid. Bidders will be expected to submit a drawing showing the various sections of the support structure. Provisions for anti-theft/vandalism should be stated.

10.5.4 Warranty, defects liability, service, and maintenance

Warranty and defects liability period. The bidder should be required to state as part of the technical proposal the warranty period and defects liability period (DLP), repairs/replacement covered by the warranty, and the extent and terms of the warranty.

After-sales support. The support expected after installation should be stated and bidders encouraged to give explicit details and commitment to providing such support in case the scheme develops a problem.

Service proposal. The bidders can be requested to provide a priced proposal with their bid for a two-year service agreement after expiry of the warranty and DLP.

The bidder should also detail as part of the technical proposal their availability and capacity to provide backup support from within the country, preferably through their physical presence in the country.

More of this is discussed in section 11.2.4.

10.5.5 Deliverables

This section defines what constitutes the indicators of project completion, which could include: a signed goods received note provided by the supplier; signed test certificate; delivery, installation, and commissioning report to the contracting agency and users; and/or a training report.

10.5.6 Bidder qualification

The criteria listed in section 10.3 should be provided to the bidders so they are aware of the requirements for qualifying as a bidder, including reputation, experience, and capacity.

10.5.7 Evaluation of tender and other considerations

Evaluation criteria. The evaluation criteria for the bids should be provided here. This will help the bidders in knowing how they will be scored so that they can provide complete and comprehensive bids meeting all requirements. Those who cannot meet these requirements will not submit a bid, saving time and effort for both parties.

Activity timeline. The activity timeline is useful for bidders to know whether they can work within the timelines given and to plan properly, thus ensuring timely delivery of the project. The bidders should also be asked to provide a work plan that fits within the stated timelines.

Obligations. The obligations of both the buying organization and the bidding organization (selected contractor) should be clearly stated, including information such as provision of site access, site handover, payment obligations, contract signing, ethical standards, logistics, delivery, and reporting.

Submission of bid. Submission guidelines should be provided to the bidders indicating the closing day and time of the tender, the submission means (email, tender box, etc.), and any other requirements, such as separation of technical and financial proposals.

Annexes. Under the annexes, documents to be attached include technical evaluation checklist, deliverables checklist, any technical drawings such as the module support structure, and a list of all equipment that should be supplied (if applicable).

A sample terms of reference document can be accessed on the Global WASH cluster under the Solar Pumping Toolkit (GLOSWI, 2018f). This sample document is provided as a simple guide that can be used to quickly and easily tailor a tender document for the specific context.

10.6 Quality of solar modules

Conformity to specified standards is one of the first, obvious indicators of quality of system components.

The quality of solar modules is worth special consideration as people often do not know how to make a choice between different modules due to so many different types being available. Besides that, many markets in developing countries are flooded with cheap, low-quality modules that are difficult to distinguish with the naked eye. Differentiating a good-quality module from a bad one is indeed difficult, yet module quality has a huge impact on system sustainability. Following is some guidance for choosing a module of good quality and performance.

10.6.1 Checking conformity and authenticity

As mentioned in section 10.2, certification is an important indicator of module quality. Certification to IEC/EN 61215, IEC/EN 61730 (for crystalline modules), IEC/EN 61646 (for thin-film modules) and UL 1703 is mandatory for any module to be considered a quality module. Modules that meet these standards provide a better assurance of prolonged life.

Optional standards can also be considered depending on the actual conditions the modules will be installed in, for example, IEC/EN 61701 is required for modules that will be used in coastal areas. This is an indicator that the module will be able to withstand the salty mist conditions of coastal installations.

Certifications are either marked on the module nameplate or on the module datasheet. Modules that are not marked with certification numbers introduce uncertainty as to their quality and genuineness.

Bidders should be asked to provide module certification identification numbers corresponding to the quoted certifications, whose genuineness can be verified on certification databases, with the applicable database depending on the certifying organization. The authenticity of the panel brand (and therefore its quality) should be called into question when bidders are unable to provide such certification identification numbers.

For example, in Figure 6.1, the SolarWorld module is certified to IEC 61215 and IEC 61730, traceable to TÜV Rheinland (the certifiying agencies for this case). The applicable database to check the authenticity of this certification is TÜV Rheinland.

A few certifying bodies and their validity-checking databases are:

- TÜV Rheinland; https://www.certipedia.com/
- TÜV SÜD Group; https://www.tuvsud.com/en/services/product-certification/ps-cert
- TÜV Rheinland; - DIN CERTCO; https://www.dincertco.tuv.com/?locale=en
- TÜV NORD; https://www.tuev-nord.de/en/company/certification/certificate-database/
- VDE Institute; https://www2.vde.com/en/Institute/OnlineService/VDE-approved-products/Pages/Online-Search.aspx

On some of these databases the certification can be searched using the company or product name.

Checking certification using company module name. Referring to the SolarWorld module in Figure 6.1:

1. Identify the certifying organization from the module datasheet. In this case the certifying organization is TÜV Rheinland.
2. Go to the certification database corresponding to the certifying organization. In this case it is <https://www.certipedia.com/>.
3. Enter the name of the module brand or the name of the manufacturer (e.g. SolarWorld) and click the 'Start Search' tab.
4. The displayed results show the SolarWorld products, including modules that are certified by TÜV Rheinland.

158 SOLAR PUMPING FOR WATER SUPPLY

Checking certification using certificate number. Trinasolar is a global solar module manufacturer. On its website are listed the different types and sizes of solar module that it makes. The datasheet for Allmax multicrystalline module 265–285 W indicates that it is IEC 61215 and IEC 61730 certified.

1. Identify the certifying organization from the module datasheet. In this case the certifying organization is TÜV Rheinland.

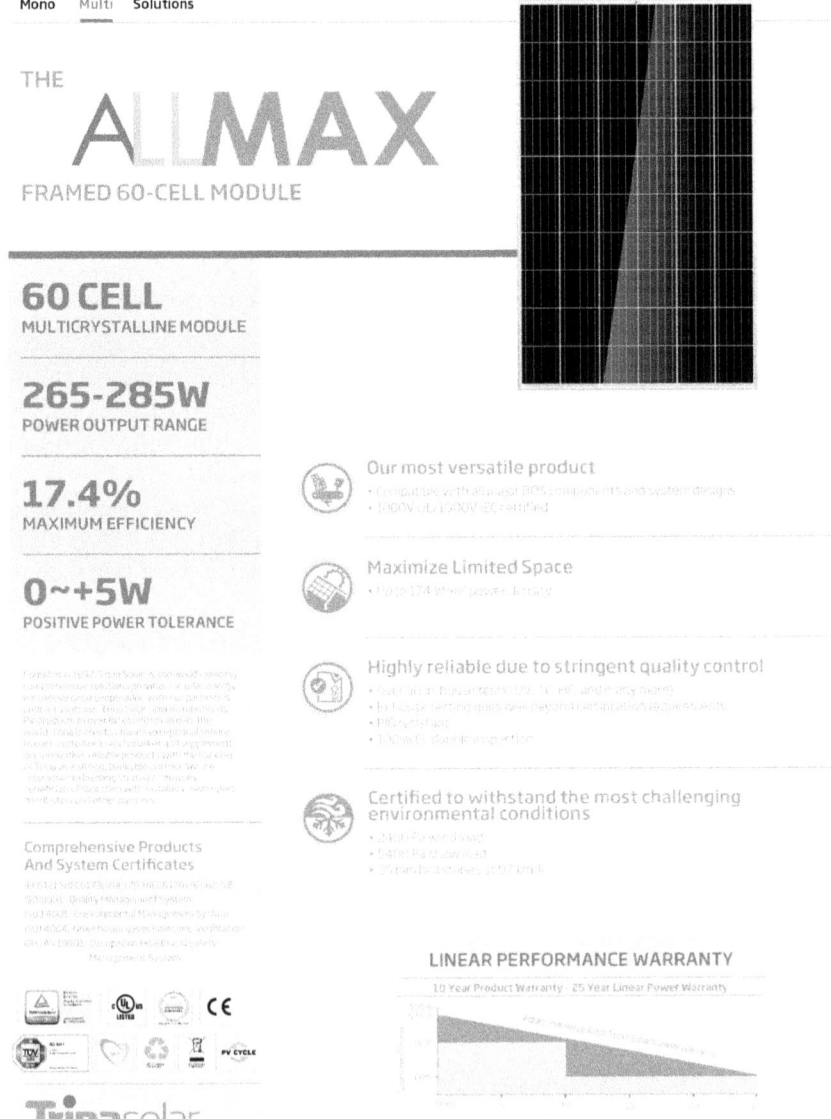

2. Identify the certification certificate ID number from the module. In this case the certificate ID is 0000024632.

3. Go to the certification database corresponding to the certifying organization: in this case it is <https://www.certipedia.com/>.
4. Enter the certification ID number and click the 'Start Search' tab.
5. The displayed results show that the certification is genuine and lists all the certified products so that the particular certified product (model type) can be verified.

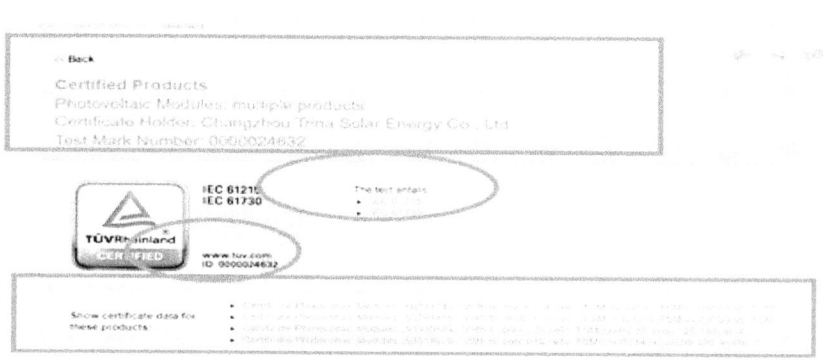

A lack of proper import controls in many developing countries has led to the flooding of markets with low-quality products and the entry of counterfeit modules. To counter this some world-leading manufacturers mark each solar module with a serial number that can be found on a sticker inside the glass of the PV module. At procurement point the serial number can be emailed to the original manufacturer for verification of the module as a genuine product from their factory. Serial numbers help to reveal counterfeits that come marked with fake stickers of leading brands.

Panels without proper manufacturing certifications with their corresponding ID number should be avoided.

10.6.2 Additional evaluation of solar module quality and performance

Additional criteria for evaluating the quality and performance of a solar module include:

Power rating. The higher the power rating of a module, the higher the power it will generate. This criterion can be used to compare different module sizes on offer.

Efficiency. A module with higher efficiency can produce more power than a similar module with lower efficiency. This criterion is particularly useful for comparing modules of the same type and size, for example, two polycrystalline modules of the same power rating but with different efficiencies.

Power tolerance. This is an indicator of how much the power output from a module in operation will differ from the power indicated on the nameplate. It is typically expressed as a plus or minus percentage. For example, a 300 Wp module with tolerance of ±5 per cent means it has a power output range of 285–315 W under STC. As a selection criterion, a smaller tolerance range is preferable as it represents a lower deviation from the actual and gives a higher certainty of getting the actual output.

> **Box 10.4 Checking module quality**
> - Certification
> - Power Rating
> - Efficiency
> - Power tolerance
> - Temperature coefficient

Temperature coefficient. The effect of cell temperature on the power output of the module is discussed in sections 2.10, 4.2 and 7.8.1. The temperature coefficient quantifies how the output from a module decreases per unit increase in cell temperature above the STC temperature of 25°C. It is expressed as a percentage having different values for power, voltage, and current. For example, a 300 Wp module with a temperature coefficient for power of –0.45 per cent/°C means for every unit increase in cell temperature above 25°C, the power output from the module reduces by 1.35 W. Modules with lower temperature coefficients are preferable because they translate into less power loss at higher temperatures.

10.7 Practical aspects of equipment and supplier selection

Many implementing agencies and organizations use limited donor funds to implement solar pumping projects. There will be a tendency to want to minimize the procurement costs often at the expense of quality. This is further exacerbated by institutional procurement procedures which require that consideration be given to the price, such that the award is given to the lowest bidder. Often engineers abdicate the responsibility of equipment selection to the procurement department which is likely not familiar with the technical aspects of the product. They will therefore give preference to price

CALLS FOR PROPOSAL AND BIDDING   161

over technical specifications under the assumption that the products are similar in specification. It is advised that engineers oversee the technical selection of products, giving priority to quality over price.

> The choice of equipment should not be left to procurement personnel without a technical review from knowledgeable engineers.

Private-sector contractors/suppliers are often constrained to provide satisfactory support when they bid at the lowest price point. While cut-throat competition has the benefit of introducing competitive prices and challenging monopolies, it could also lead to declining service levels that could compromise longevity and sustainability. This is to say that initial price should not be the sole determinant of a suitable supplier, but the support that will be provided thereafter should be used to differentiate between several suppliers. This support will likely come at a higher price. Suffice to say, goodwill as well as long-term relationships between the buyer and supplier are important for guarantee of quality and long-term support.

Another occurrence that has been observed is the direct engagement of foreign-based suppliers/manufacturers without a physical presence in the country of implementation to install the SPWS at field level. While this works at the installation stage it will usually not work in the long run. This is due to the unreachability of these suppliers in case a problem is encountered and the high costs in terms of time and money to send parts and knowledgeable technicians from abroad. Installations have been encountered that have taken months to repair, or have been abandoned altogether due to the unavailability of parts and lack of support from the international supplier. Commonly, these international suppliers will be brought in due to bilateral agreements between governments and while such engagements cannot always be avoided, measures can still be

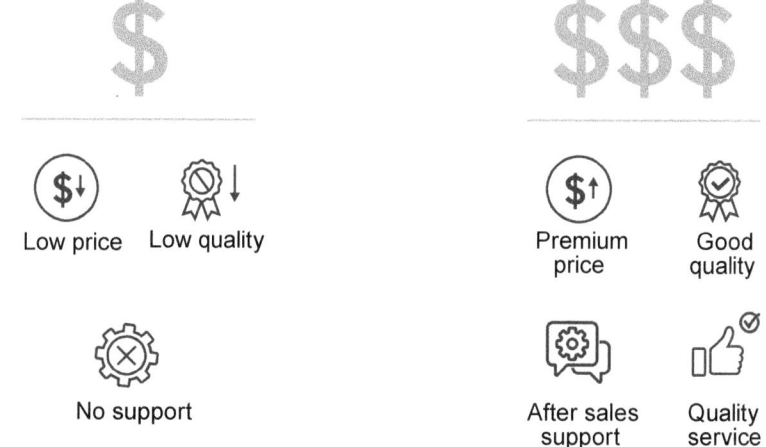

Figure 10.3 Cheap will become expensive with time

introduced to have local dealers, spare parts, and technicians in place for the future.

Finally, a single supplier policy should be favoured as opposed to buying different components from different suppliers. Besides it being easier to manage a single supplier it will add confidence to the warranty as a single supplier will take full responsibility for every component in the system without passing blame to others for failure. It is strongly discouraged to buy different products from different suppliers and engage another party for the installation in order to save costs, without considering the implications for the warranty when failure occurs.

A final point here is the prudence of using equipment that is familiar in the market, that is, equipment which is well known by local providers/technicians and which they can support with spare parts and repair. Where possible, organizations should standardize their solar pumping equipment by having one or two brands which they use across their programmes and for which they can stock spare parts or even have personnel dedicated to supporting the equipment. It is easier to institutionalize the SPWS knowledge of a few brands than multiple brands.

CHAPTER 11
Testing and commissioning, operation and maintenance

The handing over of solar pumping schemes from installers to users involves specific testing and commissioning procedures. This chapter reviews those processes together with the necessary documentation that should be provided with every scheme. The different maintenance categories, ranging from routine to preventive, reactive, predictive, and extraordinary maintenance, are explained. The importance of having dependable after-sale specialized technical support to ensure proper functioning of solar pumping schemes over time is discussed together with health and safety considerations.

Keywords: inverter programming, solar testing report, PV module cleaning, maintenance for solar systems, open-circuit voltage, short-circuit current

11.1 Testing and commissioning

11.1.1 Inspections and functionality tests

Before the solar-powered water system is put into use it should be tested by the contractor/installer for proper function and safety. The various checks and tests recommended are outlined in Table 11.1.

11.1.2 System performance tests

Once the preliminary checks have been done the working system should be tested, performance monitored, and recorded. The testing should be done for a prolonged period of two to three days over the full solar day to monitor performance, which can help in revealing any discrepancies from the expected/design values. These test results act as a benchmark for monitoring future performance. The various components of the SPWS should be subject to the tests outlined in Table 11.2.

11.1.3 Testing report

Upon satisfactory testing of the SPWS by the contractor, the contractor should prepare a testing report detailing the results of the tests and stating information on the installation as below. A detailed hourly report of the pump performance should also be provided, as shown in Table 11.3.

Table 11.1 Inspections and functionality tests for solar-powered water systems

Item/component	Description of action/test
PV generator	Visual inspection of modules for damage during installation
	General orientation of the modules
	Support structure mechanical check e.g. stability, obvious swaying
	Measurement of the rated value of modules, i.e. open-circuit current and short-circuit current at solar peak hour (when the sun is strongest)
	Disconnection switch check for polarity, insulation, cable connection, etc.
	Output voltage and current measurement at each string inside the disconnection switch at solar peak hour (a digital multimeter can be used). These should match the design values or deviate by the factor of tolerance discussed in sections 4.5.1 and 10.6.2
Inverter/control box	Visual check
	Environmental protection check
	Start-up sequence
	Maximum power point tracking
	Inverter programming
Pump and motor	Continuity test
	Insulation test
	Priming for surface pumps
Cables	Visual inspection of cables, e.g. loose connections, hanging wires, abrasions, correct dimensions, insulation
	Voltage and megger (insulation-resistance) tests
PV grounding system	Continuity of ground connection
	Measurement of ground resistance
	Protective earth tests
Pipework and fittings	Visual inspection of wellhead and pipes, e.g. loose connections, broken pipes, bent pipes, correct sizes
	Check for closed valves

11.1.4 Commissioning, acceptance, and system handover

After practical completion of the installation, the client should request commissioning, followed by acceptance, and system handover, at which stage the risk and liability passes over to the client (or implementing agency or community of users). Warranty and maintenance periods usually commence after satisfactory handover by the contractor and acceptance by the client.

Table 11.2 System performance tests for a solar-powered water system

Item/component	Description of action/test
PV generator	Measurement of the system current and voltage on load and record of PV power generated
Inverter	Output frequency, voltage, and current
	Functionality of all sensor inputs, such as low-water sensor, high-level sensor, pressure sensor, irradiation sensor
Pump and motor	Start-up sequence (reverse rotation is a common problem that can result in reduced pump output)
	Flowrate and pressure
	Motor current and speed
Cables	Obvious signs of failure, such as overheating, peculiar smells
Pipework and fittings	Check for leaking pipes and fittings, blockages, etc.

During the commissioning and handover process, the installer should provide the following documents to the client:

- signed goods received note;
- wiring diagram of the control components, including wire sizes and ratings;
- wiring diagram of the PV generator;
- operation and maintenance manual (easy to read and preferably in a local language), including troubleshooting procedures;
- technical equipment manuals, such as equipment datasheets, specific equipment operation and installation manuals;
- signed warranty document;
- capacity-building plan for the first 1 to 2 years;
- testing and commissioning report, including the testing and performance results (Table 11.3);
- other documents identified in Table 11.4.

Usually the implementing agency will in turn hand over the water scheme to the users at the end of the project implementation period. As mentioned, a complete, well thought-out and well-executed handover procedure should be conducted, involving:

- training users on basic operation and maintenance of the system;
- the provision of complete warranty information;
- the provision of all relevant documents and information regarding the operation of the system;
- linking to a technically knowledgeable third party (e.g. private contractor) to provide routine preventive and corrective maintenance.

Despite complete training and handover there are still certain aspects of the system which the users will not be able to resolve on their own if a problem

Table 11.3 SPWS testing report template

Client name/implementing agency

Installing company/contractor name, contact person, and contact details

Installation completion date

Borehole details	1. Site name 2. Location name and GPS coordinates 3. Borehole depth (m) 4. Borehole diameter (mm)	5. Water rest level (m) 6. Pumping water level (m) 7. Total dynamic head (m) 8. Depth of pump (m)
Equipment specifications	1. Pump brand and model 2. Motor brand, model, and rating 3. Motor serial number 4. Pump cable type and size 5. Piping size and type	6. Controller name, brand, and model 7. Controller serial number 8. PV module brand, model, and size 9. No. of modules in series 10. No. of modules in parallel

Pump performance	Morning							Afternoon/evening					
	0600	0700	0800	0900	1000	1100	1200	1300	1400	1500	1600	1700	1800
Irradiation (W/m^2)													
System pressure (m)													
Pump output (m^3/hr)													
Motor current (A)													
Pump speed (Hz)													
Input power (kW)													
Input voltage (VDC)													
Output voltage (VAC)													
Input current (A)													
Output current (A)													

occurs. In such cases, coordination with knowledgeable local government entities or water offices and/or technically competent private companies should be a prerequisite for project handover.

11.2 Operation and maintenance of equipment

Solar-powered water schemes will suffer fewer breakdowns and require much less intensive maintenance than generator or handpump schemes. However, solar-powered schemes can and will experience problems at some point that cannot be solved at community level (or which the managing agency/entity will probably need specialized technical support), regardless of the training provided at user level.

As such it is important that where possible, some kind of operation and maintenance (O&M) service agreements (see section 11.2.4) are established prior to any installation and renewed over the years, with a quality private contractor, water utility, water service provider, relevant government technical office, and/or any other stakeholder with sufficient technical knowledge and means to timely respond when needed.

A well-thought and professional O&M service package, ideally negotiated together with the construction contract/tender, will ensure that an SPWS maintains high levels of functionality; it will also ensure extended life of the equipment.

In addition, and especially where spare parts and knowledgeable technicians are mostly located in capital cities, as is still often the case, clustering of SPWSs in a geographical area will facilitate preventive maintenance and repairs being carried out in an easier and more effective way for everyone.

Solar-powered water schemes have been found to work without any major issues for over 10 years in communities where O&M procedures have been clear, well designed, and followed. Yet this is not often the case in humanitarian operations where time pressure tends to lead to an excessive focus on solar technology design and installation, often at the expense of a well-structured O&M plan (in terms of the plan itself and the funds dedicated to it). This puts the functionality of the solar pumping scheme at risk in the short and medium term.

Typically, technical faults might develop and repairs might be needed within the first 12 to 18 months of operation of the scheme (due often to some wrong design data or some faulty component or installation errors not detected during the short testing period prior to commissioning). Therefore, it is critical to ensure a proper O&M service programme is in place at the very least during that period, and that it is renewed and adapted to contextual situations over the years.

In the life of a solar-powered water scheme, the O&M phase is by far the longest, as shown in Figure 11.1. Therefore, increasing the quality of O&M services is important and neglecting O&M is risky, counterproductive, and more expensive in the long term.

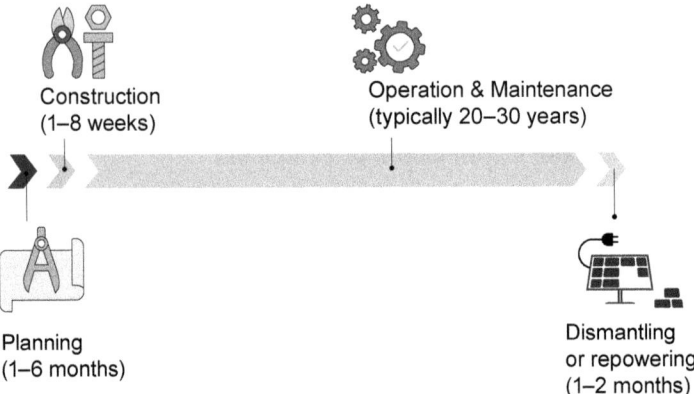

Figure 11.1 Duration of different phases in the life of a solar pumping scheme

11.2.1 Operation

Operation is about onsite daily management, supervision, and control of the solar pumping scheme (see Annex F – Routine Inspection and Maintenance Sheets) and may also involve remote monitoring (see more on remote monitoring in Chapter 12). It may also involve coordination of maintenance activities, depending on the model chosen for doing so.

SPWS documentation is crucial for an in-depth understanding of the design, configuration, and technical details of the system. It is the asset owner's responsibility to provide these documents; if not available, they should, as best practice, be recreated at the asset owner's cost (also in the local working language where appropriate).

Before assuming any maintenance and/or operational activities, it is important that the actors involved understand the characteristics of the PV scheme. Moreover, for quality/risk management and effective operations management, good and clear documentation of contract information, scheme information, maintenance activities, and PV scheme management are needed over the lifetime of the equipment.

In general, for optimum service provision and as a best practice, the contractor in charge of more technical specialized maintenance will have access to all possible documents.

The contractor or agency in charge of construction and installation of the solar pumping equipment will develop a site operating plan which gives a complete overview of the scheme location, layout, electrical diagrams, components in use, and references their operating manuals, health and safety rules for the site, and further topics as needed and agreed. This plan will have to be handed over to the party in charge of O&M and be stored (electronically and physically) safely for immediate access in case of solar pumping scheme issues.

A list of documents to be included in the as-built documentation set accompanying the SPWS is given in Table 11.4.

Table 11.4 Documentation set accompanying the solar water-pumping scheme

Item	Description
Site information	Location, GPS coordinates, detailed map
	Spare parts storage/warehouse (if any)
	Stakeholder list and contact information (e.g. maintenance and repair service provider)
Project technical studies and drawings	Plant layout and cable drawings and sizes
	PV module configuration and connection
	Earth/grounding system layout drawing
	Lightning protection system details (if any)
	Water yield simulation
	Water pump and inverter study
	List of balance-of-system components
	Drawing and location of water reservoir and pipe sizes, material, and lengths
PV modules and Inverters	PV module datasheets
	O&M inverter manual
	Inverter setting and location
	Warranties and certificates
Switchgear (disconnection switch)/Controls	Type of switchgear
	Switching procedure to operate and stop pumping system
Mounting structure	Mechanical assembly drawings and material
	Warranties and certificates
Security, anti-theft, and alarm system	Security and alarm system layout
PV scheme controls	PV scheme control system and meter descriptions
Communication system	Installation and O&M manual
	Working agreement and communication to remote-monitoring operators
Other documentation	Test and commissioning report from installation company

As best practice for the operations team, when the system is out of service and maintenance performed the maintenance tasks should be documented and linked back, if applicable, to failures that may have triggered the respective maintenance activity. This will allow the team to learn from past and ongoing operation and maintenance, and to then be able to improve performance via, for example, predictive maintenance in the following years.

Records of service and maintenance activities should include the input records listed in Table 11.5.

Table 11.5 Records of service and maintenance activities at site level

Activity type	Input record and information type
Routine onsite maintenance	List of activities and person responsible
Alarms/operation incidents	Date and time, affected equipment, number of days, external visits/inspections from third parties
Preventive maintenance	Preventive maintenance plan and activities to undertake
	Tasks performed, date, technician name, and function
Corrective maintenance and repairs	Detailed failure, problem resolution description, and start and end dates of the intervention
Inventory management	Inventory stock and management
Monitoring and supervision	Water production, inverter readings, start and stop pumping times, weather conditions
	Records of visits
Warranty management	Registration of claims

11.2.3 Maintenance

Maintenance is usually carried out onsite either by trained personnel, specialized technicians, or subcontractors. Maintenance can be of five different types, depending on the kind of activities, frequency, and level of technical skills required.

Routine maintenance. Once the solar pumping scheme has been installed and commissioned, several simple actions are to be followed by the owners of the system in order to maximize water production and minimize the likelihood of system breakdowns (typically carried out by non-specialized staff who have received some kind of training, e.g. operator, community members, NGO staff).

While these activities are simple and easy to do it is not unusual to find numerous schemes where they are not carried out either because the personnel onsite have not been trained or advised to do so, tools have not been provided, and/or the community of users and water committee are not aware of the importance of these activities and their influence on the amount of daily water supplied by the system.

> Being consistent in the performance of routine maintenance activities will strongly influence the daily amounts of water supplied.

A list of the activities involved in routine maintenance at field level can be found in Table 11.6 and in Annex F, and main ones are illustrated in Figure 11.2 (as a best practice, a similar poster to this can be placed by the water point to act as a reminder over the years).

Preventive maintenance. A core element of maintenance services, preventive maintenance involves regular visual and physical inspections as well as verification activities to comply with the operating manuals.

TESTING AND COMMISSIONING, OPERATION AND MAINTENANCE   171

Table 11.6 Routine maintenance activities for solar-powered water systems

	Activity	Frequency
Solar pumping maintenance	PV module cleaning	Weekly to monthly, depending on the level of dust
	Vegetation trimming and management	As needed to avoid shadows on PV modules
	Maintenance of PV supporting structure	As needed and monthly check of bolting of panels to the structure
	Inspect water piping, repair or report leakages	Weekly to monthly
General site maintenance	Pest control, waste disposal	Weekly
	Perimeter fencing repair	As needed
	Maintenance of buildings	As needed
	Maintenance of security equipment (if any)	Test functionality once a month
Onsite measurement	Water readings	Daily
	Recording of pumping hours, weather conditions	Daily
	Inverter readings, fault recording and reporting	Daily

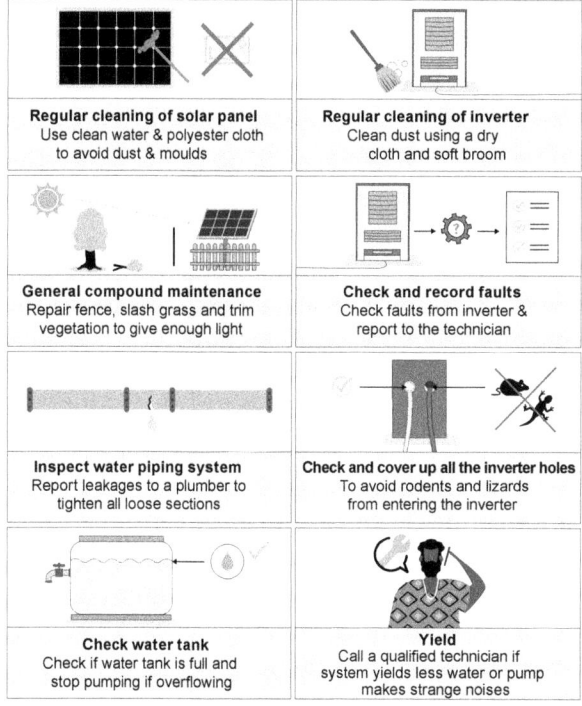

Figure 11.2 Routine maintenance activities to ensure proper daily functionality

> **Application note: suggested method for cleaning PV modules**
>
> Cleaning PV modules is perhaps the single most important regular maintenance activity to be carried out. For proper cleaning, use only water (as pure and absent of suspended solids as possible) and cloth; no soap or other elements are needed. Pressure washing modules is not an acceptable cleaning method and may invalidate the warranty on the panels.
>
> For panels mounted on elevated poles or roofs, cleaning is made easier when a ladder and a telescopic handle is provided. Health and safety must be considered to keep workers safe onsite. This will involve some form of specific training for solar module cleaning, including working at height if cleaning elevated mounted modules.

The preventive maintenance plan details a list of inspections that should be performed at predetermined intervals (typically quarterly, biannually, or annually depending on context) by a technician with specialized knowledge (e.g. technical staff from a private contractor). Tracking records of preventive maintenance carried out will optimize activities further. (See Annex G for a preventive maintenance plan template.)

The maintenance contract (section 11.2.4) should include this scope of services and each task frequency. Ideally, such a contract will be negotiated together with (and even be an integral part of) the installation contract.

It is the responsibility of the contractor in charge of maintenance to prepare the preventive maintenance plan for the duration of the contract period, following the frequencies specified in the contract. These activities should be reported to the asset owner or asset manager.

Corrective maintenance. Corrective maintenance covers activities aimed at restoring a faulty solar pumping scheme, equipment, or component to a status where it can perform the required function. Agreeing beforehand on maximum response time to minimize downtime of the solar pumping scheme should be stated in the O&M service agreement. A system of financial penalties could be put in place to ensure that technicians will be deployed onsite when called in a reasonable amount of time (one to three days).

When a system falls into disrepair it is critical that non-qualified staff do not tamper with the equipment as this can aggravate the problem and/or lead to electric shocks or other accidents as well as void the warranty. In any case, modifications should not be carried out on the solar pumping scheme without technical guidance from the system supplier, the contractor in charge of maintenance, or a qualified technician. Box 11.1 describes the scenarios when a technician should be called.

> **Box 11.1 When to call a technician**
>
> - When the pump is making unusual noises.
> - When there is any change in the rate of pumping (the system is pumping less water than it used to yet the solar modules are clean).
> - When biannual maintenance checks need to be performed.

Corrective maintenance takes place after a failure is detected either by onsite operators and users, remote monitoring and supervision staff, or during regular inspections and specific measurement activities. Operators should be clear about the occasions when it is necessary to call a qualified technician onsite.

Only when a qualified technician has arrived should corrective maintenance take place, which will involve the following three activities:

1. fault diagnosis, also called troubleshooting, to identify the cause of the fault and isolate it;
2. temporary repair to restore the required faulty item for a limited time, until a full repair is carried out;
3. full repair to restore the required function permanently.

In cases where the PV plant or segments need to be taken offline, the execution of scheduled corrective maintenance during the night or low irradiation hours would be considered best practice as the overall power generation is not affected.

Predictive maintenance. Predictive maintenance is a special service provided by contractors in charge of maintenance who follow best practice principles. It is defined as a condition-based maintenance carried out following a forecast derived from the analysis and evaluation of the significant parameters of the solar-powered water scheme.

A prerequisite for good predictive maintenance is that information can be retrieved from the devices onsite in such a way that the contractor in charge of maintenance can evaluate trends or events that signal deterioration of the different devices, especially the PV modules or water pumps.

For efficient predictive maintenance a certain level of maturity and experience is required, which is at best a combination of knowledge of the system's performance, related equipment design, operation behaviour, and relevant accumulated experience and track record from the service provider.

Normally, it is a process that starts after the implementation of an appropriate monitoring system and the creation of a baseline. The baseline will represent the entire SPWS operation as well as how different pieces of equipment interact with each other, and how this system reacts to environmental changes.

Although predictive maintenance is rare to find in the humanitarian context, it has several advantages, including:

- maintenance activities (both corrective and preventive) are anticipated;
- some maintenance activities can be delayed, eliminated, or optimized;
- time for repairs is reduced;
- spare parts replacement costs are reduced;
- emergency and non-planned work is reduced;
- predictability is improved.

Extraordinary maintenance. Extraordinary maintenance, usually not covered by the O&M service agreement and fee, may be necessary after major unpredictable events occur in the solar pumping scheme that require substantial repair works (e.g. damage after theft, fire, vandalism, or caused by design mistakes).

11.2.4 Service contract framework

One of the key challenges facing SPW systems is the lack of locally available technical expertise to provide troubleshooting, repairs, and specialized maintenance services to the users. This is more so once the after-sales service period has ended, where all liability for service and maintenance shifts to the user/owner of the system. This commonly also corresponds with the time when the system is being handed over to the community/users by the implementing entity. This results in uncertainty of getting the help needed in case of system failure which could lead to prolonged periods of system downtime and non-functionality.

A service contract plays a key part in tackling this challenge. It is signed between the users and the installer (or other qualified maintenance contractor) for a period of one to two years post system handover to the users or within the critical first 18–24 months of the system's operation. It will be an agreement with specific terms (including annual fees) between the asset owner and the contractor in charge of maintenance. This agreement defines in detail the maintenance services, both remote operations services and onsite maintenance activities, the management and interface of those services, and the responsibilities of each party. Liquidated damages, penalties, and bonus schemes are also part of the contractual commitments.

> **Example 11.1 Preventive maintenance service contract with Solar Pumping Ltd in Kenya**
>
> Under service agreements with Solar Pumping Ltd the private contractor is mandated to visit the system at least four times a year to conduct regular maintenance checks. They are also required to visit the system when a problem arises.
>
> Clients in the service contract are charged annually. Consumers are charged based on their distance from the nearest Solar Pumping Ltd branch, at $0.90 per kilometre, for each routine check and a technician cost of $100 per day.
>
> In case of equipment breakdown, the client pays for repair/replacement of components at a discounted rate. This expense is in addition to the annual service and maintenance fee.

It is considered best practice to define and agree a maintenance framework before starting construction of the SPWS. As some installers will void the equipment warranty if they are not engaged to provide the maintenance service, it is advisable to consider the repercussions of engaging a different maintenance contractor from the original system installer. The agreement is renewable on an annual basis upon both parties meeting their obligations.

> **Box 11.2 Example of essential provisions in a maintenance service agreement**
>
> **Communication line**
>
> - Reporting to asset owner on solar pumping scheme performance, O&M performance, incidents, and warranty management
>
> **Solar pumping scheme operation**
>
> - Scheme documentation management
> - Scheme operation and/or supervision
> - Performance monitoring and documentation
> - Issue detection/diagnostics
> - Security incidents monitoring interface
>
> **Solar pumping scheme maintenance**
>
> - Maintenance scheduling
> - Solar pumping scheme and site maintenance activities, and contact details for preventive and corrective maintenance
>
> **Spare parts management**
>
> - Spare parts list for two to five years
> - Spare parts replenishment
> - Spare parts storage (optional)

The scope of services to be provided by the contractor in charge of maintenance should include:

- quarterly, biannual (twice a year) or annual checking up and record-keeping for proper operation of the system;
- scheduled/routine visits by the contractor to site according to the agreement (commonly every two or three months) to carry out preventive maintenance, address impending failures before they happen (such as replacing overheating breakers that could cause inverter or pump failure, checking tightness of connections), replace any components or subcomponents of the SPWS as needed, and ensure optimal operation of the system;
- prompt response by the contractor to address any reported faults by the client, with turnaround time for such repairs not exceeding three days (or as stipulated in the service agreement);
- continuous training of the users/operators by the installer on the operation and maintenance of the system;
- submission to the client of a report of the works carried out during every visit and the state of the system, including performance (flow, pressure) and electrical data (current, voltage, power).

In most cases the service contract fee covers the cost of labour and transport for carrying out such works with the replacement parts being paid for by the client.

One challenge of putting in place such a plan is the shortage of funds, though donors can support this by extending funding windows to cover one to two years of maintenance fees after project handover. The implementing body should put in its funding proposal the maintenance cost for one to two years after system handover.

This kind of maintenance plan is beneficial as it contributes towards overall success, functionality, and sustainability of the system. It provides a link between the community/users and the private sector so that the community is not left in limbo and is guaranteed of system functionality in the long run. Further, on expiry of the initial NGO-funded maintenance agreement, the community should be aware of its value and be encouraged to renew the contract using water fee collections or other funding sources.

Suffice to say, a robust maintenance plan is absolutely essential for a successful and sustainable SPWS. This includes but is not limited to an operation and maintenance plan, a service contract, and continuous capacity-building of the users.

11.2.5 Training

Training of operators and other personnel (e.g. water committee members, NGO technical staff) in the running and daily maintenance of the solar pumping system is critical.

As shown in the previous section, it is strongly recommended that this training be a contractual condition so that it is provided by the installation contractor before leaving the site and on an ongoing basis according to the contractual agreements.

Components of the training may cover all relevant aspects of the O&M of the solar plant as well as basic plumbing (useful in repairing leakages) and the daily running of the water scheme, including being able to manage finances accrued from water sales if relevant.

As solar pumping solutions are rapidly expanding in many countries, NGOs, UN agencies, and other parties should consider utilizing the technical expertise of private contractors to provide regular technical and awareness training. Given the business opportunity, some contractors may offer such a service for free.

Finally, everyone who enters a solar pumping scheme site, regardless of their skills and experience, needs to be trained in the dangers present in addition to the individual skills and experience that are required for the tasks they normally perform. Awareness of the necessary health and safety regulations is a must.

11.2.6 Health and safety

Solar pumping schemes are electricity-generating power stations and pose significant hazards which can result in permanent injury or death. It needs to be remembered that PV modules cannot be switched off and as long as there is sunlight, even if the pump is not in operation, electrical shock hazards are present. Risks can be mitigated through proper hazard identification, careful planning of works, regular briefing of procedures to be followed, and well-documented inspection and maintenance practices.

The dangers of electricity are well known and can be effectively managed through properly controlled access and supervision by the contractor in charge of maintenance. Any person accessing an SPWS location should be briefed on hazards and risks.

Staff working on electrical equipment must be appropriately trained, experienced, and supervised, but it is also key that others working around the equipment, for example module cleaners, are equally aware of the potential risks and have safe methods of working.

Hazardous areas and equipment should carry appropriate signage to warn personnel of the hazards and wiring sequence. Such markings should be clear and evident to all personnel and third parties (and intruders) entering the scheme premises.

Besides workers of the solar plant, it is not unusual for other parties to require access to it. This may be the asset owner, or their representative, the landlord, or in some situations members of the public. It is important that the plant access control and security system keeps people away from areas of danger and that they are appropriately supervised and inducted as necessary.

The asset owner is ultimately responsible for compliance with health and safety regulations within the site/plant.

11.2.7 Security

It is important that the solar pumping scheme site, or key areas of it, are protected from unauthorized access. This serves the dual purpose of protecting the equipment (from theft or vandalism) and keeping members of the public safe.

Together with the contractor in charge of maintenance and the security service provider, the asset owner will put in place a security protocol in case an intrusion is detected.

A security system may be formed of simple fencing or barriers, or may include other measures as listed in section 7.6.

As well as the general security of the site over the lifetime of the scheme, particular attention should be paid to periods of construction or maintenance when usual access arrangements may be different. It is important that security is always maintained, particularly when there are activities that may be of more interest to members of the public, children, or potential thieves.

11.2.8 Spare parts management

Spare parts management can be an inherent and substantial part of O&M that should ensure spare parts are available in a timely manner for corrective maintenance in order to minimize the downtime of a solar-powered water scheme (or a part of it).

Generally, it is not economically feasible to stock spare parts for every possible failure in the water scheme. Therefore, the contractor in charge of maintenance together with the asset owner should define a list and stocking level of specific spare parts that make economic sense.

CHAPTER 12
Warranties, social models for management, and monitoring

This chapter describes warranties related to products, installation, and services for solar pumping schemes. As important as the technical considerations, social models of management adapted to the specificities of solar pumping schemes are essential to ensure its functioning in the long term; the different models are presented here together with commonalities in well-managed schemes. Finally, remote operation and maintenance is briefly explained, along with the different range of tools to monitor electrical parameters in a solar PV generator.

Keywords: solar component warranty, electrical monitoring tools, solar key performance indicators, remote monitoring, thermal image, solar checker

12.1 Warranties

Reputable equipment manufacturers provide product warranties for their equipment. The supplier of the equipment must provide to the asset owner a warranty document clearly detailing and explaining the terms of warranty, including the start date and the period of warranty (see Annex E).

The terms generally include repair, replacement, or refund of salvage value of the defective equipment due to failures caused by manufacturer fault. Warranty does not cover failure that is due to external causes.

The warranty period starts from the time of either collection, delivery of equipment or commissioning, in the case of installation by the manufacturer's local representative/distributor/supply agent/installer. The warranty is claimed through the supplier/installer, who should extend the manufacturer's warranty to the buyer or end user.

> Warranty is voided when installation instructions are not followed and when repairs are attempted without the manufacturer's authorization.

Often the installation contractor will combine the *manufacturer's component warranty* with their own warranty terms to include *workmanship, quality of installation* and *after-sales support* under what is known as a *comprehensive warranty*. The warranty provided should therefore cover both the product and workmanship. Proof of purchase must be provided for a warranty claim to be honoured.

Incomplete handover procedures (such as when the community/users/committees are not provided with warranty documents) can create issues if there are problems with the system within the critical two-year mortality period (a system that lasts the first two years of operation has a much greater probability of lifetime functionality). It is therefore paramount that during the handover process the users are educated on the warranty claim process, provided with warranty documents, linked with the right contact in case of a problem, shown where to access technicians, and given all relevant information regarding the warranty. Field evaluations have revealed systems that have remained non-functional for long periods of time (within the warranty period) with obvious consequences, and which could have been quickly and easily rectified had the users known the avenues for warranty and repair. Provision of warranty certificates to the users and education on warranty procedures helps to avoid such occurrences.

12.1.1 Warranty management

The technical asset manager or the asset owner will be focal points for any warranty claims to the contractor or manufacturers of SPWS components.

During the warranty period (or at least the first two years), it should be mandatory, as far as possible, for the contractor to enter into a comprehensive maintenance contract with the asset owners, and this should be well stated in the contract documents.

For any warranty claims the formal procedure provided by the warranty provider should be followed; therefore, it is important that warranty terms and conditions are made clear and are well understood at the time of equipment purchase and installation.

All malfunctions or failures of equipment should be reported to the manufacturer or contractor promptly as the warranty validity is considered at the date of reporting, not when the problem was observed.

The warranties typically requested by implementing agencies are component, quality of installation and workmanship, and system performance warranties.

Component warranty. The length of the warranty period varies with the type of equipment and the manufacturer. The standard warranties provided by the manufacturers of the individual components, usually against defective components or workmanship, are:

- *Solar modules*; 10-year product warranty and 25 years power output guarantee based on light-induced degradation (see section 4.5.3);
- *Controller/inverters*; 2–5 years;
- *Pump and motor*; 2–5 years;
- *All other balance of system components*; 1–2 years.

Unless the warranty specifies that these are covered onsite (i.e. within the defects period, or under an extended maintenance contract), then they are assumed to be on an exchange basis and the labour and costs of transport are excluded, that is, the equipment must be returned by the client to the supplier/contractor for diagnosis of the problem. In case the installer needs to visit the site for diagnosis, the cost of the visit is charged to the client, although this also depends on the goodwill extended by the supplier and other contractual obligations stipulated in the contract. Where it is found that the failure was due to the fault of the installer, all costs are borne by the installer. This can be a point of contention which should be discussed and documented during the contract signing/negotiation stage. It is advisable to look for more than individual component warranties.

Quality of installation and workmanship warranty. The minimum defects liability period or after-sales service support commits the installer to provide backup support onsite (typically within the first six months to one year), without additional charge, in case the system develops a problem after installation. It requires that any items which fail or are not installed to standard are corrected and resolved onsite at cost to the supplier. Issues arising from poor quality of installation are also dealt with onsite at a cost to the supplier and this also covers corrosion of materials used onsite, such as the support structure.

An after-sales warranty follows the same principle as the component warranty in that only failures that are caused by contractor fault, workmanship, or quality issues are covered.

System performance warranty. Although rarely encountered in humanitarian contexts, the supplier may guarantee that the PV system will meet or exceed the design performance for several years.

A system performance warranty may be based more generally on monthly average or annual average figures for insolation and water delivery, or on actual instantaneous outputs. Performance can be assessed initially during commissioning via the system performance test benchmarking and the supplier should warranty that future system performance tests, when corrected back to design conditions, will exceed the design performance.

As long-term system performance is subject to many onsite and environmental conditions as well as requisite maintenance, such warranties would need to go hand in hand with a long-term maintenance contract.

12.1.2 Activities typically covered and not covered by warranties

During the warranty, the following maintenance will be required to be carried out by the contractor/supplier/installer:

1. repair/replacement of all defective components and subcomponents of the system as per the requirement to ensure the system is in good working order;
2. repair/replacement of the SPWS to make the system functional within the warranty period whenever a complaint is lodged by the user. The contractor shall attend within a reasonable period of time

and, in case of breakdown, fix issues within a period not exceeding three days of being reported;
3. repair/replacement of components damaged due to negligence or fault of the client (users and/or asset owner), theft, or vandalism, at the clients' cost.

The safety and security of the system will be the sole responsibility of the user/asset owner.

12.2 Social models for management

As discussed in section 11.2, operation and maintenance of solar water systems is crucial to their long-term sustainability and inadequate O&M is a frequent cause of system failure.

When an NGO or UN agency takes over the full management of a solar-powered water scheme, as is often the case in camp settings, management (including O&M) becomes simple and straightforward because it is assumed that those organizations will make provisions in their budgets and keep in contact with competent technical partners to respond to any issues regarding the operation and functionality of the water scheme.

For solar pumping schemes at community level things are normally more complex. Different management models work in different settings and the authors do not advocate for or discourage any particular model. Different studies have tried to analyse the strengths and weaknesses of a number of management models at community level (see for example WSTF, 2017). However, because a model is working in one setting does not mean it will work in all settings; and that a model is not working in one setting does not mean it's a bad idea for another. Good concepts may sometimes not work because they are either not executed well, their execution was wrongly timed, or they are not suited to the particular environment.

Several evaluations and field visits carried out within the Global Solar and Water Initiative have found that there are three key factors which should be incorporated into any management model to ensure its sustainability:

- financial accountability of water committees collecting water fees (e.g. legally registered water committees, with an open bank account, keeping registers of contributions);
- dedicated personnel (e.g. salaried water operators);
- involvement of government technical offices and private companies for repairs and services.

Other commonalities observed in successful community-managed water schemes are described in Box 12.1.

Financial sustainability models found at community level for solar-powered water schemes are like those found for other technologies, namely: pay as you go, metered monthly payments, and flat monthly payments.

> **Box 12.1 Commonalities between successful community-managed water schemes in Kenya**
>
> **At least one individual who is highly committed to the success of the project**
> For instance, in Adamasija, Wajir County, the committee chairman has taken it upon himself to ensure that meters are read, has set up a shop where people can settle their water bills, ensures that electricity bills are paid, and oversees the system's water distribution schedule.
>
> **Absence of alternative sources of water**
> Lack of alternatives, especially during the dry season, causes communities to be better stewards of their water systems. This may mean a higher willingness to pay for water to ensure availability of funds for maintenance. This is especially the case where the main source of livelihoods (e.g. livestock) is highly dependent on availability of water.
>
> **High sense of community responsibility**
> A general perception that water collections by the committee are community funds. Some of these committees had contributed towards the construction of classrooms in schools within their communities, extended distribution networks to key institutions, or provided water at subsidized costs during the dry season.
>
> **Community awareness of recurrent costs**
> Communities with diesel or solar–diesel hybrid systems were seen to have a higher understanding of financial management, which is key to the sustainability of water schemes. These schemes also saw that consumers had a higher sense of willingness to pay for water services, unlike in stand-alone solar systems, where there was a general wrong perception that water should be free as the energy is free. See also Box 12.2.
>
> *Source*: GLOSWI, 2018f

> **Box 12.2 For local community projects, should solar water be free?**
>
> Given the nature of implementation of solar-powered water systems for communities, which are often through aid or government projects (at no cost to the benefitting communities), and as there is no cost for the energy required for water pumping, this tends to be interpreted as 'water for free', thus negating creating a perception that water should not be paid for.
>
> There needs to be a complete shift in the community narrative of SPWSs, from 'tapping into a cost-free source of energy to pump water' to 'cumulating funds for system replacement'. Communities need to understand that while there are minimal recurrent costs in operating solar systems, there are significant one-off costs to ensuring their continued long-term operation. For long-term sustainability of systems, water must be provided at a fee to ensure funds for repairs, maintenance, and replacement of parts, and implementing agencies need to be very deliberate in altering this viewpoint.

12.3 Monitoring

12.3.1 Remote monitoring

Monitoring on a system-wide basis is very important; it can help to further develop the installation, evaluate overall performance, and reduce costs.

The main things to be monitored at the solar pumping side of a water scheme are:

- discharge from the borehole, amount of water pumped;
- operating pressure;
- static and dynamic water level;
- pump operation trends e.g. current, frequency;
- solar radiation on the array plane, voltage, and current.

Solar pumping technology allows monitoring of those parameters either onsite with the use of specific tools or through remote monitoring solutions (see Figure 12.1).

Remote monitoring allows for distant, computer-based operation and monitoring of a solarized water scheme in areas where a telephone network is present. In humanitarian operations, remote monitoring is useful to run and monitor water schemes that are distant from the humanitarian organization's operational base, when schemes are many and staff is stretched, where tight controls are needed (e.g. critical boreholes supplying a large number of people), and/or where security or other logistic constraints make it difficult to access the scheme when required.

In addition, remote monitoring is useful to better understand aquifer behaviour and anticipate maintenance needs through recording and observing actual and historical data. Actual data collected may be live monitoring, system status, 'pump off' reasons, and advanced information (e.g. faults), while historical data includes running time; water amounts, and static and dynamic levels; voltage, current, flow, and so on.

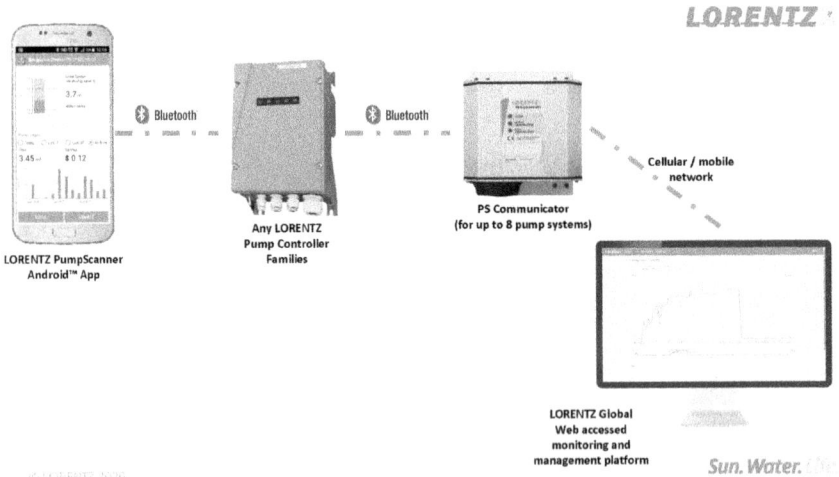

Figure 12.1 Lorentz communication system for monitoring from phone via Bluetooth or from computer via Internet
Source: Lorentz

12.3.2 Key performance indicators

Key performance indicators should reflect the performance of the SPWS and assess the O&M service provided by the O&M contractor.

SPWS KPIs include:

- water supplied vs the water forecast;
- performance ratio (the energy generated divided by the energy obtainable under ideal conditions expressed as a percentage);
- uptime/availability, parameters that represent, as a percentage, the time during which the plant operates over the total time it is possible to operate. While uptime reflects all downtime regardless of the cause, availability involves certain exclusion factors to account for downtime not attributable to the O&M contractor (such as force majeure), an important difference for contractual purposes.

O&M contractor KPIs may include:

- acknowledgement time (the time between a problem being reported and the contractor acknowledging it);
- intervention time (the time between acknowledgement and a technician reaching the SPWS site);
- resolution time (the time to resolve the fault, starting from the moment the technician reaches the PV plant);
- response time (acknowledgement time plus intervention time), an indicator used for contractual guarantees.

12.3.3 Monitoring and diagnosis tools

Clamp meter. Unless a more specialized monitoring and maintenance service for a considerable number of solar pumping stations is required as simple clamp meter of the right category (see Figure 12.2), able to measure both voltage and current (AC and DC) to the required scale, should be enough to perform basic electrical checking.

Solar meter. Solar radiation levels in the panel can be easily measured by placing a solar meter on top of a panel (in order to ensure the same orientation and tilt angle). By measuring solar radiation levels and water supplied and comparing them with expected values according to the design provided it is possible to get an idea of the accuracy of the design and outputs expected and act in consequence.

Thermography. Thermography cameras are used to spot high temperature points and cells in panels with a temperature difference of more than 15 °C (normally indicating some kind of defect or problem). Thermography can also be used to locate bad connections in cable boxes or inverters.

Other problems that can be typically spotted with thermal imagery are:

- welding defects;
- impurities in silicon cells of panels;

Figure 12.2 Fluke clamp meter

- defects in connection box of solar modules;
- reverse polarization of solar cells;
- loose cable connection in inverter/control boxes;
- cooling or overload issues at any solar equipment part.

Solar checker. Solar checkers are a more complete, complex, and expensive testing and monitoring device that can be used to provide PV panel, array, or whole generator measurements onsite. Especially used to trace I-V curves for output measurement and spotting differences with expected values, these tools are useful to locate potential problems at solar generator level but need specialized training for their use.

ANNEX A
Pump and generator design basics

This annex provides a quick guide for determining a duty point for pump design together with the design of a diesel generator.

Two components are required to select a suitable pump: the design yield (m³/hr) and the total head. These two parameters define the duty point of the required pump.

Design yield

There are two ways of determining the design yield.

1. Based on the daily water needs of the population to be served. The daily demand is converted to an hourly flow by dividing by the hours of pumping. (For stand-alone SPWSs, the hours of pumping are equivalent to the peak sun hours but for hybrid systems this is the total number of hours of combined pumping.)

$$\text{Design yield } (m^3/hr) = \frac{\text{Daily demand } (m^3/day)}{\text{Hours of operation}}$$

2. Based on extraction of the maximum potential of the borehole. In this case the design yield is matched to the safe yield (see section 5.3.2) of the borehole so that the maximum potential of the borehole is pumped.

$$\text{Design yield } (m^3/hr) = \text{Safe yield } (m^3/hr)$$

In both cases, the design yield should never exceed the safe yield of the borehole (60–70 per cent of the tested yield).

Total head calculation

For boreholes,

$$\text{Total dynamic head (TDH)} = H_{DWL} + H_{elev} + H_{friction} + H_{residual}$$

Where,

H_{DWL} = dynamic water level, i.e. the water level after completion of test pumping measured from the borehole surface

H_{elev} = elevation/vertical height from borehole surface/wellhead to the tank inlet

188 SOLAR PUMPING FOR WATER SUPPLY

$H_{friction}$ = friction loss, i.e. pressure drop due to friction in the pipe expressed as a coefficient of friction per 100 m. The coefficient of friction is obtained from friction loss tables

$H_{residual}$ = residual head, i.e. additional pressure required at the delivery point. For pumping to tank it is a value between 0 m and 10 m

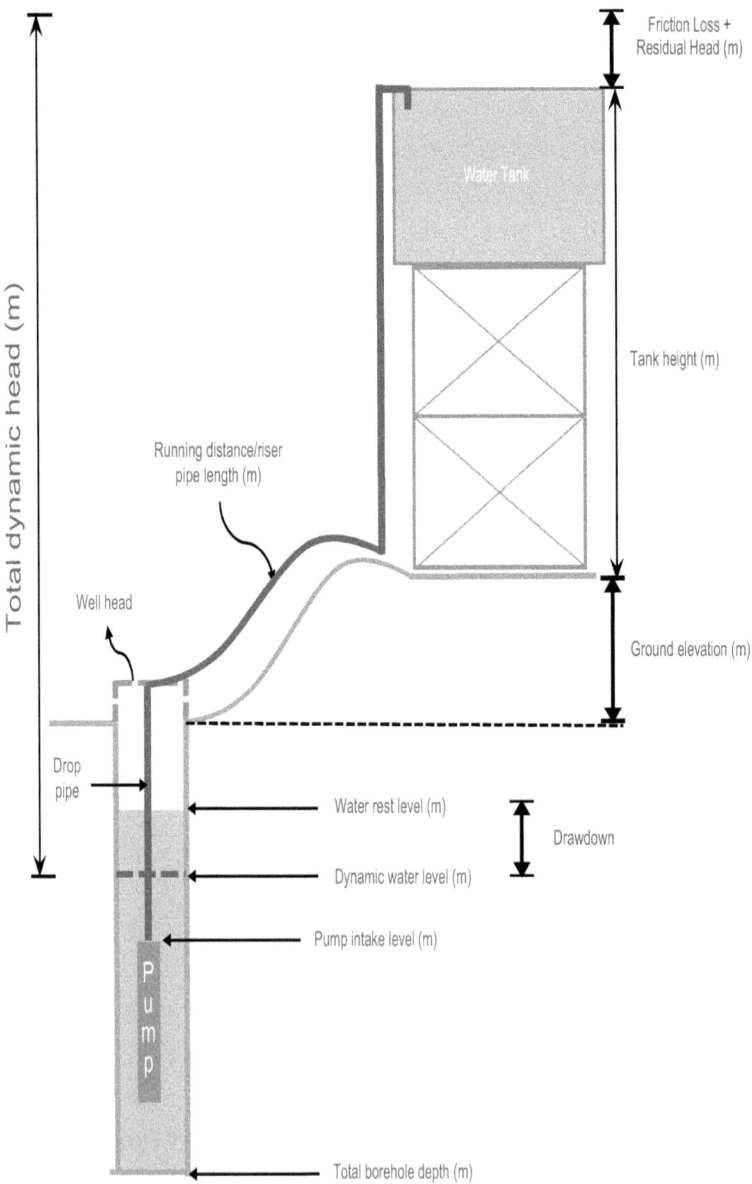

Figure A1 Layout of total dynamic head calculation

For surface pumps,

$$\text{Total head (TD)} = H_{suction} + H_{static} + H_{friction} + H_{residual}$$

Where,
- $H_{suction}$ = vertical height from water level to pump inlet where the water level is below the pump inlet, i.e. negative suction. Where the water level is above the pump inlet, this is a positive suction
- H_{static} = elevation/vertical height from pump inlet to the tank inlet
- $H_{friction}$ = friction loss, i.e. pressure drop due to friction in the pipe expressed as a coefficient of friction per 100 m. The coefficient of friction is obtained from friction loss tables
- $H_{residual}$ = residual head, i.e. additional pressure required at the delivery point. For pumping to tank it is a value between 0 m and 10 m

Pressure loss due to friction loss is affected by various factors:

1. Length of the pipe – the longer the pipe, the greater the pressure drop will be due to friction.
2. Diameter of the pipe – the smaller the pipe, the greater the pressure drop.
3. Flow of water through the pipe – the greater the flow, the greater the pressure drop.
4. Pipe roughness – the rougher the interior surface of the pipe, the greater the pressure drop. PVC pipe is smoother than galvanized iron pipe and has less friction in comparison.
5. Fittings and joints – each bend, elbow, union, flow meter, valve, strainer etc. introduces an additional pressure drop that must be considered.

All these factors increase the resistance to the flow of water through the pipe, thus increasing the pressure drop leading to higher pumping power requirements.

Pipe friction is determined by using friction factors (coefficients of friction), which are available in pipe friction tables. To use these tables the following must be known: the flow to be pumped through the pipe, the total running distance of the pipe, the inner diameter of the pipe, the pipe type (whether PVC or GI) and the pipe class (class C, class D etc.). The friction tables give head loss per 100 m which is then worked out for the entire length, as explained in the worked example below.

The residual head refers to the additional pressure requirement at the delivery point. For delivery to a tank the value can be zero. For delivery to an irrigation scheme the residual head will be the pressure requirement of the irrigation nozzle/head (refer to Chapter 8, Box 8.1).

For surface pumps, besides the total head, due consideration should also be given to the suction capability of the pump under the conditions in which it is installed. This suction capability is called the maximum suction lift of the pump. Lorentz provides a good step-by-step process for calculating the maximum suction lift (https://partnernet.lorentz.de/files/lorentz_psk2-cs_manual_en.pdf). In other sources this maximum suction lift is also explained using the net positive suction head (NPSH) required verses the NPSH available.

Generator sizing

It is common for WASH engineers to size the generator by multiplying the pump motor size by a factor of two to three. This section seeks to explain why this is done with the aim that engineers will pay better attention to correct sizing and avoid the common problem of oversizing generators, which leads to higher fuel consumption and running costs.

Engine-driven generators are tested, rated, and offered by manufacturers according to standard conditions of:

- maximum altitude above sea level of 150 m;
- maximum air inlet temperature of 30 °C;
- maximum humidity of 60 per cent.

Operating generators in conditions that exceed these results in reduced power from the same generator. Generators will often be installed in locations that are outside the standard conditions, which leads to power losses because of higher altitude, higher temperature, and higher humidity.

What this means is that to determine how much power the generator will produce (or working in reverse, what size of generator should be selected to meet the power demand of the pump), deration of the generator has to be done using the power loss values provided here:

- altitude: loss of 3.5 per cent for every additional 300 m above the maximum recommended 150 m above sea level (2.5 per cent for turbo-charged engines);
- air inlet temperature: loss of 2 per cent for every 5.5 °C above 30 °C (3 per cent for turbo-charged engines);
- humidity: loss of 6 per cent at 100 per cent humidity.

In addition, the efficiency of the alternator is taken into consideration together with voltage reduction during start-up.

The formula for calculating the generator size is:

Generator size (kVA) = Motor size × efficiency factor × start-up factor
× deration for altitude × deration for temperature
× deration for humidity × 1.25

Worked example

Parameter	Value
Borehole depth	118
Tested yield	13 m³/h
Water rest level	31
Dynamic water level	41
Safe yield	7.8–9.1 m³/hr
Pump intake level	93 m, using 2-inch steel pipe
Distance to tank/running distance	1350 m, using 2-inch PVC pipe, class E

PUMP AND GENERATOR DESIGN BASICS 191

Parameter	Value
Ground elevation	3 m
Tank height	6 m
Number of elbows in the pipeline	3No. x 63 mm
Number of gate valves	1No. x 2 inch
Number of non-return valves	1No. x 2 inch
Location and altitude	3°N 38°E, 540 m
Temperature and Humidity	38°C, 80 per cent

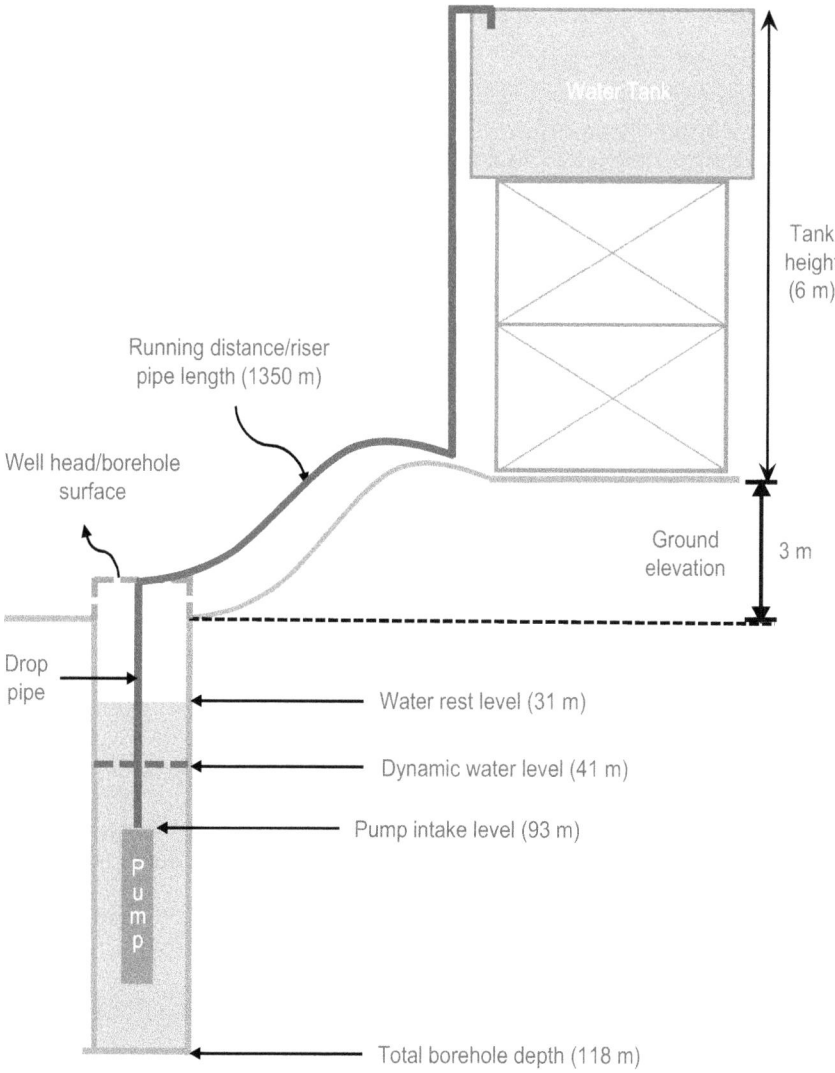

Figure A2 Layout of worked example

Design yield

Based on maximum extraction of the borehole's potential, the design yield = safe yield = 7.8–9.1 m³/hr

TDH calculation

First we need to determine the length of pipe equivalent to the fittings in the system as these also introduce losses into the system. The equivalent length of pipe for the various 2-inch fittings and valves can be computed using tables available from different sources such as https://powderprocess.net/Tools_html/Piping/Pressure_Drop_Key_Piping_Elements.html or https://www.engineeringtoolbox.com/resistance-equivalent-length-d_192.html

3 elbows: 3 × 3.0 = 9 m
1 gate valve: 0.6 m
1 non-return valve: 5.2 m
Total equivalent length = 14.8 m

Total effective riser pipe length for friction computation is therefore: 1350 m + 14.8 m = 1364.8 m

Friction tables are used to pick the corresponding coefficient of friction (such as the one available here titled TABLE 1: PVC AND GI FRICTION LOSS TABLE https://dayliff.com/media/com_hikashop/upload/safe/technical_reference.pdf)

- The coefficient of friction for a flow of 9 m³/hr through a 2-inch steel drop pipe is 4.7 m per 100 m.
- The coefficient of friction for a flow of 9 m³/hr through a 2-inch PVC class E riser pipe is 3.4 m per 100 m.

$$\text{Therefore, total friction loss} = (4.7/100 \times 93) + (3.4/100 \times 1364.8)$$
$$= 4.37\,\text{m} + 46.38\,\text{m} = 50.77\,\text{m}$$

$$\text{Total dynamic head} = H_{DWL} + H_{elev} + H_{friction} + H_{residual}$$
$$= 41 + (6 + 3) + 50.77 + 10 = \mathbf{110.77\ m}$$

Pump selection

From the design yield and TDH determined above, the duty point for this system is **9 m³/hr at 110 m head**. This is used to select a suitable pump.

Referring to the pump performance curves in Figure B4 (in Annex B), a submersible borehole pump (a Grundfos SP9-21 4.0 kW 3 × 400 V 50Hz) can be used to meet this duty point.

Generator selection

To size a generator for the **4 kW** pump, the following steps should be followed.

Step 1: Allow for losses due to efficiency of alternator *(80 per cent)*

$$= 4\ \text{kW} \div 0.8 = \mathbf{5\ kW}$$

Step 2: Oversize for voltage reduction during start-up *(35 per cent)*

$$= 5 \text{ kW} \times 1.35 = \mathbf{6.75\ kW}$$

Step 3: Compute deration factors
Altitude: 3.5 per cent for every 300 m above 150 m above sea level

$$\frac{540 - 150}{300} \times 3.5\% = 0.0455\ (4.55\%)$$

Temperature: 2 per cent for every 5.5 °C above 30 °C

$$\frac{38 - 30}{5.5} \times 2.0\% = 0.0290\ (2.9\%)$$

Humidity: 6 per cent at 100 per cent humidity
At 80 per cent humidity, deration factor is 3 per cent

Step 4: Apply the deration factors to get the generator size

$$\frac{6.75 \text{ kW}}{(1 - 0.0455) \times (1 - 0.0290) \times (1 - 0.03)} = \frac{6.75}{0.9545 \times 0.971 \times 0.97} = \mathbf{7.508\ kW}$$

Step 5: Convert kW to KVA (Cosq = 0.8)

$$\mathbf{7.508 \times 1.25 = 9.385\ kVA}$$

A generator that is at least 9.385 kVA is required. In this case select the next nearest size that can be found in the market i.e. 10 kVA.

ANNEX B
Manual calculation of solar system

Chapter 5 explained why it is preferable to use computer-based software to design the PV generator of a water scheme over manual design. While the manual process does not produce precise results and cannot accurately predict the actual performance of the SPWS, it is useful in cases where computer-based software is not available, as with some solar pumping brands that do not provide design software for sizing.

The method for selecting a pump discussed in Annex A results in a pump brand and model, with a factory-defined motor size. In other scenarios, there could be an existing pump with a known motor rating. The motor needs to be powered to drive the pump that delivers water to the tank or to the point of use. In the case of an SPWS, the motor is powered using electricity generated by the PV generator. The process of estimating the energy to be generated by the PV generator is discussed below. This process details how many solar modules should be installed to power the pump motor to deliver the desired pump performance. It should be kept in mind that the process can result in either an undersized or oversized PV generator, leading to suboptimal performance of the system.

For any pump, the power absorbed by the motor from the power source, P_1, can be calculated in either of two ways.

1. Obtained from pump curves

$$P_1 = \text{Shaft power, } P_2 \div \text{Motor efficiency, } \eta_m \quad \text{(i)}$$

2. Calculated using the duty point (flow and head)
   First, Shaft power, P_2 = Hydraulic power, P_h ÷ Pump efficiency, η_p and hence;

$$P_1 = \text{Hydraulic power, } P_h \div \text{Pump efficiency, } \eta_p \div \text{Motor efficiency, } \eta_m \quad \text{(ii)}$$

Shaft power is the power transferred from the motor to the shaft of the pump (power absorbed by the pump shaft at a given flow rate) and depends on the efficiency of the pump. The efficiency of the pump is specified by the pump manufacturer.

Hydraulic power is the theoretical power transferred from the pump to the water.

$$\text{Hydraulic power, } P_h = \frac{Q \times \rho \times g \times H}{(3.6 \times 10^6)} \text{ kW}$$

Where, Q = flow capacity (m³/h)
ρ = density of fluid (1000 kg/m³)
g = gravity (9.81 m/s²)
H = total head (m)

Total energy demand of the pump is given by:

$$E_{pump} = P_1 \times \text{Hours of operation} \qquad \text{(iii)}$$

From chapter 4, the energy generated by the installed PV generator is:

$$E_{generated} = P_{peak} \times PSH \times PR \qquad \text{(iv)}$$

Where,

- P_{peak} is the peak power of the PV generator at STC (obtained as the product of the PV module's peak power multiplied by the number of modules used)
- PSH (peak sun hours) is the equivalent number of hours per day when solar irradiance averages 1000 W/m².
- PR (performance ratio) quantifies the reduction in solar energy generated due to system losses.

Equating the two equations (iii) and (iv) gives $E_{pump} = E_{generated}$ and hence,

$$P_1 \times \text{Hours of operation} = P_{peak} \times PSH \times PR \qquad \text{(v)}$$

The performance ratio (PR) is discussed in detail in Chapter 4 and includes all the losses in the system.

Once the power of the pump (P_1), hours of operation on solar, the PSH in the location and an estimated PR are known, it will be possible to calculate the P_{peak} (or the PV generator size to be installed).

$$P_{Peak} = \frac{P_1 \times \text{hours of operation}}{PSH \times PR} \qquad \text{(vi)}$$

Since P_1, PSH and PR are all constant, it means that increasing the PV generator size increases the hours of operation. It should, however, be noted that the energy provided by solar is only available during sunlight hours. This means that the solar PV generator cannot be increased infinitely to improve the hours of pumping. As mentioned in sections 3.2.1 and 5.3.8, pumping beyond the solar day should be done using an alternative power source, such as grid or diesel power.

From the peak power, the quantity of modules required based on the models available in the market can be determined.

$$\text{Quantity of modules, } P = \frac{P_{Peak}}{\text{Module rating}} \qquad \text{(vii)}$$

Once the number of modules has been determined, their arrangement in series and parallel is then determined. The inverter to be used influences the series/parallel configuration.

The selection of an appropriate inverter is based on the pump selected and the PV generator size. The inverter should be able to handle the incoming DC power from the PV generator and condition it to that which is required by the pump.

The system voltage of the PV system is defined by the inverter voltage. Every inverter has a maximum input voltage (V_{max_input}) and a minimum MPPT voltage ($V_{min-mppt}$) which is provided on the inverter data sheet. The inverter will also have a maximum current and power input.

In addition, every solar module has a maximum voltage (V_{oc}) and peak voltage (V_{mp}) which is marked on the module and used to determine how many modules will be connected in series to be within the minimum and maximum system voltage.

If M = the number of modules in series (one series arrangement is commonly referred to as a string)
N = the number of strings in parallel

Then,

$$\text{Total quantity of modules, } P = M \times N \text{ (see Figure B1)} \quad \text{(viii)}$$

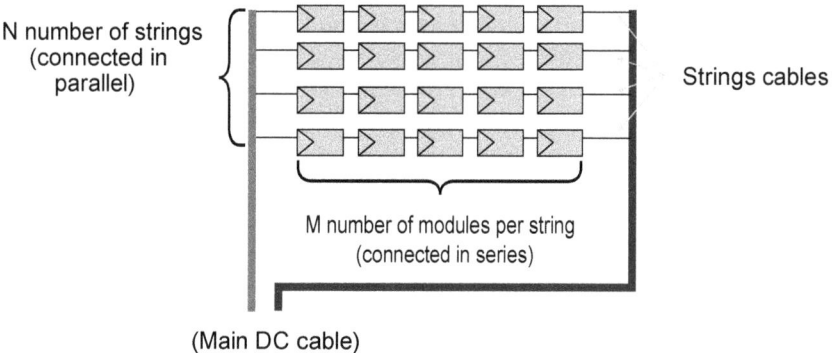

Figure B1 Number of modules arranged in series and parallel

The number of modules in series (M) is determined using the following limits.

$$\frac{V_{min\_mppt}(\text{system})}{V_{mp}(\text{panel})} < M < \frac{V_{max\_input}(\text{system})}{V_{oc}(\text{panel})} \quad \text{(ix)}$$

The value of M should be as big as possible to provide the maximum voltage possible. In other words, the value of M chosen should be as close as possible to the upper limit.

Since P and M are now known, the value of N (number of parallel strings) can be determined by plugging into formula viii.

Worked example

Step 1: Determine the water demand and calculate the required duty point (design yield and head)

Determination of duty point is explained in Annex A. Assuming the water demand is 60 m³/day, and using a duty point of 8.8 m³/hr at 110 m head, it means that the pump must be operated for approximately **7 hours** at peak pumping to meet the demand using solar alone.

Step 2: Select a suitable pump for the duty point

Using a submersible borehole pump, a Grundfos SP9-21 4.0 Kw 3 × 400 V 50Hz can be used to meet this duty point. See pump head-flow curve (Figure B4) and pump data (Figure B5) below.

Step 3: Determine the pump power demand, P_1

From the P_2/Q curves (see power-flow curve below in Figure B6) at 8.8 m³/hr the power absorbed by the pump shaft is $P_2 = 3.77$ kW.

Grundfos SP9-21 is fitted to a 4 kW motor and so the loading on the motor is 3.8/4 = 95 per cent. The efficiency of the motor at this loading level can be obtained from the motor data sheet (Figure B7) by interpolating between $\eta_m = 78.4$ per cent (at 75 per cent loading) and $\eta_m = 78.0$ per cent (at 100 per cent loading). Interpolating between the two values yields: at 95 per cent loading, motor efficiency, $\eta_m = 78.08$ per cent.

Note: For simplicity, the motor efficiency at the nearest loading (in this case 78.0 per cent) can also be used instead of interpolating.

Therefore, power absorbed by the motor from formula (i) is

$$P_1 = P_2 \div \eta_m \text{ which is } 3.77 \div 0.7808 = \mathbf{4.828 \text{ kW}}$$

Or alternatively using formulae (ii) and referring to Figure B3 below, $\eta_p = 0.7$

$$P_1 = P_h \div \eta_p \div \eta_m = (8.8 \times 9.8 \times 1000 \times 110) \div (3.6 \times 10^6 \times 0.7 \times 0.7808)$$
$$= 4.821 \text{ kW (close to 4.828)}$$

Step 4: Determine the available solar resource

Assuming the installation will be done in Sub-Saharan Africa (1ºN 39ºE in Kenya, see irradiation map in Figure B8), the average solar irradiation is 2190 kWh/m² (equivalent to **6 peak sun hours**) (source: Solargis, Kenya solar resource map).

MANUAL CALCULATION OF SOLAR SYSTEM   199

98699059 SP 9-21

Input - summary

Water volume (max): 60 m³/day
Month for sizing: June
Static lift above ground: 110 m
Dynamic water level: 0 m
Sun tracking: No (fixed)
Location: 603, Sericho, Isiolo, Kenya
Latitude: 1 DD, Longitude: 39 DD

Products

Pump: SP 9-21, 1 x 98699059
Solar module: 36 x Trinasolar 270
Switch box / control unit: RSI 3x380-440V IP66 5.5kW 12A, 1 x 99044351
Switch box / control unit: OTDCP16, Circuit Breaker, 16Amp, 2 x 98341686
Switch box / control unit: OVR PV 40-1000 P, Surge Protection, 1 x 98341687
Others: Sine-wave filter, 1 x 96754976

Sizing results - summary

Water production, Peak flow and Price
Total water production per year: 24500 m³
Avg. water production per day: 67.2 m³/day
Average water production per watt per day: 6.9 l/Wp/day

Solar module configuration:
Number of solar modules in series: 18, in parallel: 2
Solar array rated power: 9.72 kW
Solar array rated volts: 556.2 V
Sun tracking: No (fixed)
Tilt angle: 15 deg.

Typical performance at solar radiation 800 W/m²
Flow: 8.3 m³/h
Total head: 110.0 m

Cables and pipes:
Pump cable length: 120 m
Pump cable size: 4 mm²
Total cable loss: 2.2 %

Pipe diameter: DN50(44)
Friction loss: 0.0 m

System performance - monthly average

	Jan	Feb	Mar	Apr	May	Jun	Jul	Aug	Sep	Oct	Nov	Dec
Water production [m³/day]	69.6	70.1	68.4	66.5	64.8	62.6	63.3	67	69.6	69.4	67	68.2
Energy production [kWh/day]	44.2	44.5	43.4	42.3	41.2	40.1	40.5	42.5	44.0	44.0	42.7	43.4
Radiation horizontal [kWh/m² day]	6.8	7.4	7.1	6.9	7.0	6.6	6.6	7.2	7.7	7.2	6.1	6.2
Radiation tilt [kWh/m² day]	7.2	7.6	6.9	6.4	6.1	5.7	5.8	6.5	7.4	7.2	6.4	6.6
Avg. Temp. [K]	28.	29.	30.	28.	28.	27.	26.	26.	27.	28.	27.	27.

Solar data location: Latitude: 1 DD, Longitude: 39 DD

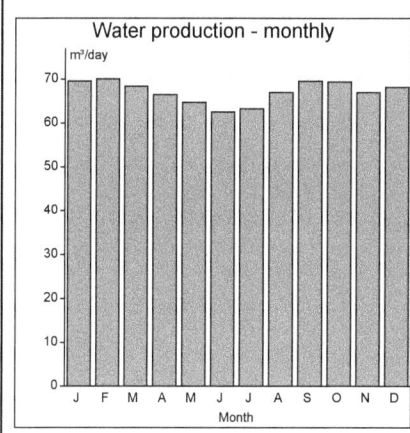
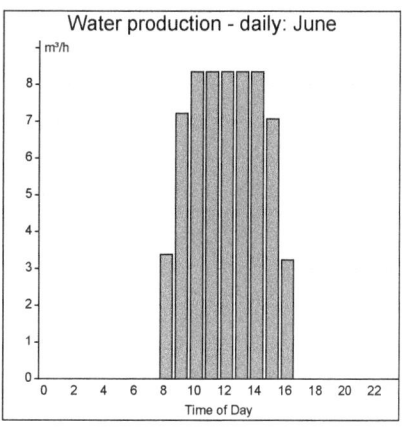

Printed from Grundfos Product Centre [2019.08.002]

Figure B2 Grundfos sizing result

 BERNT LORENTZ GmbH

Wednesday, 03 July 2019
New project

Solar pumping project

Parameter

Location:	, (1° North; 39° East)	Water temperature:	25 °C		
Required daily output:	60 m³; Sizing for June	Dirt loss:	5.0 %	Motor cable:	120 m
Pipe type:	-	Total dynamic head:	110 m	Pipe length:	-

Products

	Quantity	Details
PSk2-7 C-SJ8-30	1 pc.	Submersible pump system including controller with DataModule, motor and pump end
Trinasolar 270	40 pc.	10,800 Wp; 20 x 2 modules; 15 ° tilted
Motor cable	120 m	6 mm² 3-phase cable for power and 1-phase cable for ground
Accessories	1 set	Surge Protector, PV Disconnect 1000-40-5, PV Protect 1000-125, SmartPSUk2, SmartStart, Well Probe V2

SunSwitch setting in PumpScanner min. 200 W/m²
Daily output in June 61 m³

Daily values

	Jan	Feb	Mar	Apr	May	Jun	Jul	Aug	Sep	Oct	Nov	Dec	Av.
Output [m³]	82	83	75	67	65	61	62	68	76	75	68	73	71
Energy [kWh]	59	60	54	48	47	44	45	49	55	54	50	53	52
Irradiation [kWh/m²]	6.6	6.8	6.1	5.4	5.3	4.9	5.0	5.5	6.2	6.0	5.4	5.8	5.7
Rainfall [mm]	0.37	0.27	1.2	3.2	1.2	0.10	0.10	0.067	0.13	1.3	2.4	1.1	0.93
Ambient temp. [°C]	28	29	29	27	27	27	27	26	26	27	26	27	27

Hourly values

	6:00	7:00	8:00	9:00	10:00	11:00	12:00	13:00	14:00	15:00	16:00	17:00	18:00
Output [m³/h]	0	0	3.8	6.2	7.9	8.6	8.9	8.5	7.6	5.9	3.5	0	0
Energy [kWh]	0	1.2	2.9	4.2	5.3	5.8	6.0	5.7	5.1	4.0	2.7	1.2	0
Irradiation [kWh/m²]	0	0.12	0.29	0.45	0.58	0.65	0.68	0.65	0.58	0.45	0.29	0.12	0
Ambient temp. [°C]	22	22	23	25	27	29	31	32	32	32	32	31	31

1/5 Created by LORENTZ COMPASS 3.1.0.135
All specifications and information are given with good intent, errors are possible and products may be subject to change without notice.

Figure B3 Lorentz sizing result

SP 9

Performance curves

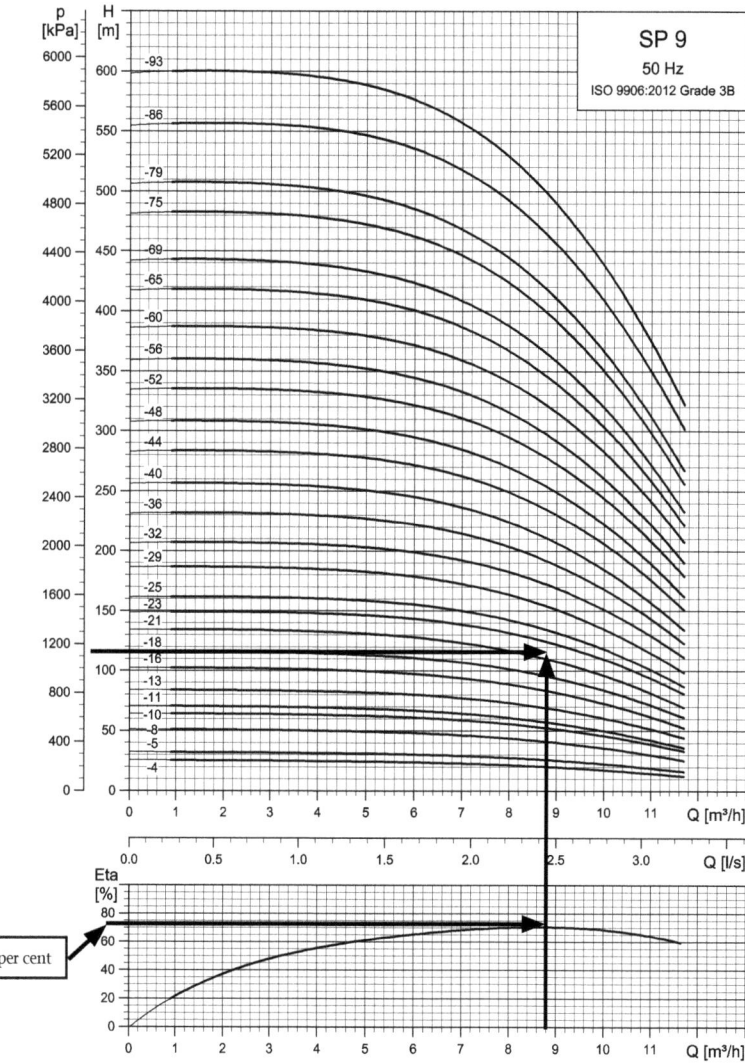

NPSH: Minimum inlet pressure 0.5 m.

Figure B4 Grundfos SP 9 head-flow curve

Dimensions and weights

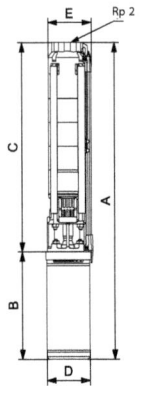
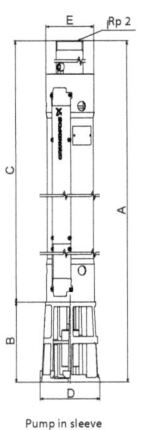

Pump in sleeve

Pump type	Motor Type	Motor Power [kW]	C	B	A	D	E	Net weight [kg]
Single-phase, 1 x 230 V / 1 x 240 V									
SP 9-4	MS 402	0.75	438	306	744	95	101	15.9
SP 9-5	MS 402	1.1	488	346	834	95	101	18.3
SP 9-8	MS 402	1.5	638	346	984	95	101	20.0
SP 9-10	MS 4000	2.2	738	577	1315	95	101	31.6
SP 9-11	MS 4000	2.2	788	577	1365	95	101	32.2
Three-phase, 3 x 220-230 V / 3 x 380-400-415 V									
SP 9-4	MS 402	0.75	438	276	714	95	101	14.7
SP 9-5	MS 402	1.1	488	306	794	95	101	16.5
SP 9-8	MS 402	1.5	638	346	984	95	101	20.0
SP 9-10	MS 402	2.2	738	346	1084	95	101	22.5
SP 9-11	MS 402	2.2	788	346	1134	95	101	23.1
SP 9-4	MS 4000	0.75	438	402	840	95	101	19.2
SP 9-5	MS 4000	1.1	488	417	905	95	101	20.7
SP 9-8	MS 4000	1.5	638	417	1055	95	101	22.5
SP 9-10	MS 4000	2.2	738	457	1195	95	101	25.6
SP 9-11	MS 4000	2.2	788	457	1245	95	101	26.2
SP 9-13	MS 4000	3	888	497	1385	95	101	29.3
SP 9-16	MS 4000	3	1038	497	1535	95	101	31.0
SP 9-18	MS 4000	4	1138	577	1715	95	101	36.2
SP 9-21	MS 4000	4	1288	577	1865	95	101	37.9
SP 9-23	MS 4000	5.5	1388	677	2065	95	101	44.1
SP 9-25	MS 4000	5.5	1488	677	2165	95	101	45.2
SP 9-29	MS 4000	5.5	1688	677	2365	95	101	47.7
SP 9-32	MS 4000	7.5	1838	777	2615	95	101	53.4
SP 9-36	MS 4000	7.5	2038	777	2815	95	101	55.7
SP 9-40	MS 4000	7.5	2238	777	3015	95	101	58.0
SP 9-23	MS 6000	5.5	1451	547	1998	139.5	139.5	55.0
SP 9-25	MS 6000	5.5	1551	547	2098	139.5	139.5	562
SP 9-29	MS 6000	5.5	1751	547	2298	139.5	139.5	58.6
SP 9-32	MS 6000	7.5	1901	577	2478	139.5	139.5	63.4
SP-9-36	MS 6000	7.5	2101	577	2678	139.5	139.5	65.8
SP-9-40	MS 6000	7.5	2301	577	2878	139.5	139.5	68.1
SP 9-44	MS 6000	9.2	2501	607	3108	139.5	139.5	78.2
SP 9-48	MS 6000	9.2	2701	607	3308	139.5	139.5	80.6
SP 9-52	MS 6000	11	2901	637	3538	139.5	139.5	86.1
SP 9-56 [1]	MS 6000	11	3396	637	4033	139.5	140	110.0
SP 9-60 [1]	MS 6000	13	3596	667	4263	139.5	140	116.5
SP 9-65 [1]	MS 6000	13	3846	667	4513	139.5	140	120.9
SP 9-69 [1]	MS 6000	13	4046	667	4713	139.5	140	124.3
SP 9-75 [1]	MS 6000	15	4346	702	5048	139.5	140	133.6
SP 9-79 [1]	MS 6000	15	4546	702	5248	139.5	140	137.1
SP 9-86 [1]	MS 6000	18.5	4896	757	5653	139.5	140	147.6
SP 9-93 [1]	MS 6000	18.5	5246	757	6003	139.5	140	153.7

[1] SP 9-56 to SP 9-86 are mounted in sleeve for R2 connection.

The pump types above are also availabl e in N- and R-versions. See page 6.

E = Maximum diameter of pump inclusive of cable guard and motor.

Figure B5 Grundfos SP 9 pump data sheet

MANUAL CALCULATION OF SOLAR SYSTEM **203**

Power curves

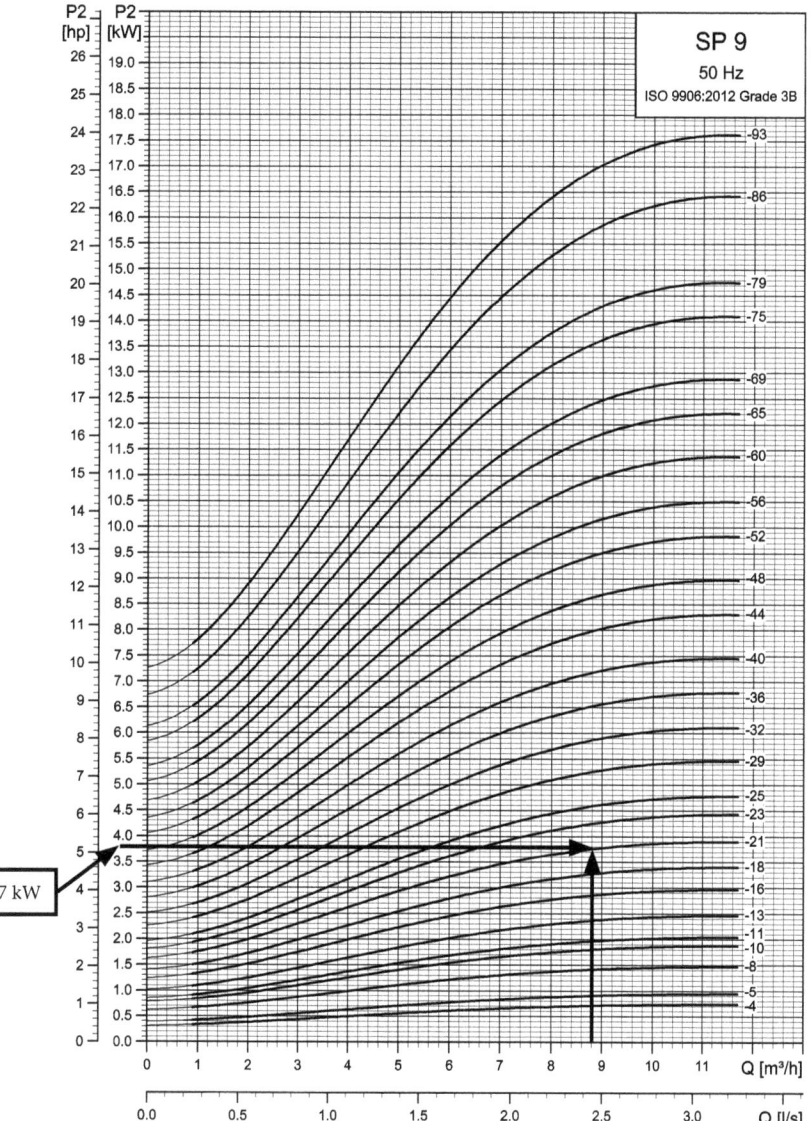

Figure B6 Grundfos SP 9 power-flow curves

3 x 230 V, submersible rewindable motors "MMS"

Motor			Full-load current I_n [A]	Electrical data						I_{st}/I_n	Dimensions		Weight [kg]
				Motor efficiency [%]			Power factor				Diameter [mm]	Build in length [mm]	
Type	Size	Power [kW]		η50 %	η75 %	η100 %	Cos φ 50 %	Cos φ 75 %	Cos φ 100 %				
MMS 6 (N, R)	6"	5.5	25.0	71	75	76	0.61	0.72	0.78	3.5	144	807	50
MMS 6 (N, R)	6"	7.5	33.5	72	76	77	0.59	0.71	0.78	3.5	144	837	53
MMS 6 (N, R)	6"	9.2	40.5	74	77	78	0.59	0.71	0.78	3.6	144	867	55
MMS 6 (N, R)	6"	11	50.0	74	78	79	0.53	0.66	0.74	3.8	144	897	60
MMS 6 (N, R)	6"	13	56.0	77	80	80	0.57	0.69	0.77	3.9	144	927	65
MMS 6 (N, R)	6"	15	62.5	79	82	82	0.58	0.71	0.79	4.3	144	997	77
MMS 6 (N, R)	6"	18.5	75.0	80	82	82	0.61	0.75	0.81	4.2	144	1057	83
MMS 6 (N, R)	6"	22	87.0	82	84	83	0.61	0.74	0.81	5.3	144	1087	95
MMS 6 (N, R)	6"	26	106	81	83	83	0.57	0.7	0.78	5.6	144	1157	105
MMS 6 (N, R)	6"	30	118	82	83	82	0.63	0.76	0.82	4.8	144	1212	110
MMS 6 (N, R)	6"	37	148	82	84	83	0.59	0.72	0.81	5.4	144	1312	120
MMS 8000 (N, R)	8"	22	82.5	80	84	84	0.71	0.80	0.84	5.3	192	1010	126
MMS 8000 (N, R)	8"	26	95.5	81	84	84	0.76	0.83	0.86	5.1	192	1050	134
MMS 8000 (N, R)	8"	30	110	83	85	86	0.71	0.80	0.84	5.7	192	1110	146
MMS 8000 (N, R)	8"	37	134	83	86	86	0.73	0.82	0.85	5.7	192	1160	156
MMS 8000 (N, R)	8"	45	168	84	87	88	0.62	0.74	0.81	6.0	192	1270	177
MMS 8000 (N, R)	8"	55	214	84	87	88	0.57	0.70	0.77	5.9	192	1350	192
MMS 8000 (N, R)	8"	63	210	87	89	89	0.81	0.87	0.90	5.7	192	1490	218
MMS 10000 (N, R)	10"	75	270	84	86	86	0.72	0.81	0.85	5.4	237	1500	330
MMS 10000 (N, R)	10"	92	345	83	85	86	0.65	0.77	0.82	5.6	237	1690	385
MMS 10000 (N, R)	10"	110	385	85	86	86	0.80	0.86	0.88	5.7	237	1870	435

3 x 400 V, submersible motors "MS"

Motor			Full-load current I_n [A]	Electrical data						I_{st}/I_n	Dimensions		Weight [kg]
				Motor efficiency [%]			Power factor				Diameter [mm]	Build in length [mm]	
Type	Size	Power [kW]		η50 %	η75 %	η100 %	Cos φ 50 %	Cos φ 75 %	Cos φ 100 %				
MS 402	4"	0.37	1.40	51.0	59.5	64.0	0.44	0.55	0.64	3.7	95	229	5.5
MS 402	4"	0.55	2.20	48.5	57.0	64.0	0.42	0.52	0.64	3.5	95	244	6.3
MS 402	4"	0.75	2.30	64.0	69.5	73.0	0.50	0.62	0.72	4.7	95	279	7.7
MS 4000R	4"	0.75	1.84	68.1	71.6	72.8	0.69	0.79	0.84	4.9	95	401	13.0
MS 402	4"	1.1	3.40	62.5	69.0	73.0	0.47	0.59	0.72	4.6	95	309	8.9
MS 4000R	4"	1.1	2.75	70.3	74.0	74.4	0.62	0.74	0.82	5.1	95	416	14.0
MS 402	4"	1.5	4.20	68.0	73.0	75.0	0.50	0.64	0.75	5.0	95	349	10.5
MS 4000R	4"	1.5	4.00	69.1	72.7	73.7	0.55	0.69	0.78	4.3	95	416	14.0
MS 402	4"	2.2	5.50	72.5	75.5	76.0	0.56	0.71	0.82	4.7	95	349	11.9
MS 4000 (R)	4"	2.2	6.05	67.9	73.1	74.5	0.49	0.63	0.74	4.5	95	456	16.0
MS 4000 (R)	4"	3.0	7.85	71.5	74.5	75.2	0.53	0.67	0.77	4.5	95	496	17.0
MS 4000 (R)	4"	4.0	9.60	77.3	78.4	78.0	0.57	0.71	0.80	4.8	95	576	21.0
MS 4000 (R)	4"	5.5	13.0	78.5	80.1	79.8	0.57	0.72	0.81	4.9	95	676	26.0
MS 4000 (R)	4"	7.5	18.8	75.2	78.2	78.2	0.52	0.67	0.78	4.5	95	777	31.0
MS 6000 (R)	6"	5.5	13.6	78.0	80.0	80.5	0.55	0.67	0.77	4.4	139.5	547	35.5
MS 6000 (R)	6"	7.5	17.6	81.5	82.0	82.0	0.60	0.73	0.80	4.3	139.5	577	37.0
MS 6000 (R)	6"	9.2	21.8	78.0	80.0	79.5	0.61	0.73	0.81	4.6	139.5	607	42.5
MS 6000 (R)	6"	11	24.8	82.0	83.0	82.5	0.65	0.77	0.83	4.7	139.5	637	45.5
MS 6000 (R)	6"	13	30.0	82.5	83.5	82.0	0.62	0.74	0.81	4.6	139.5	667	48.5
MS 6000 (R)	6"	15	34.0	82.0	83.5	83.5	0.64	0.76	0.82	5.0	139.5	702	52.5
MS 6000 (R)	6"	18.5	42.0	83.5	84.5	83.5	0.62	0.73	0.81	5.1	139.5	757	58.0
MS 6000 (R)	6"	22	48.0	84.5	85.0	83.5	0.67	0.77	0.84	5.0	139.5	817	64.0
MS 6000 (R)	6"	26	57.0	84.5	85.0	84.0	0.66	0.77	0.84	4.9	139.5	877	69.5
MS 6000 (R)	6"	30	66.5	84.5	85.0	84.0	0.64	0.77	0.83	4.9	139.5	947	77.5

Figure B7 Grundfos motor data sheet

MANUAL CALCULATION OF SOLAR SYSTEM 205

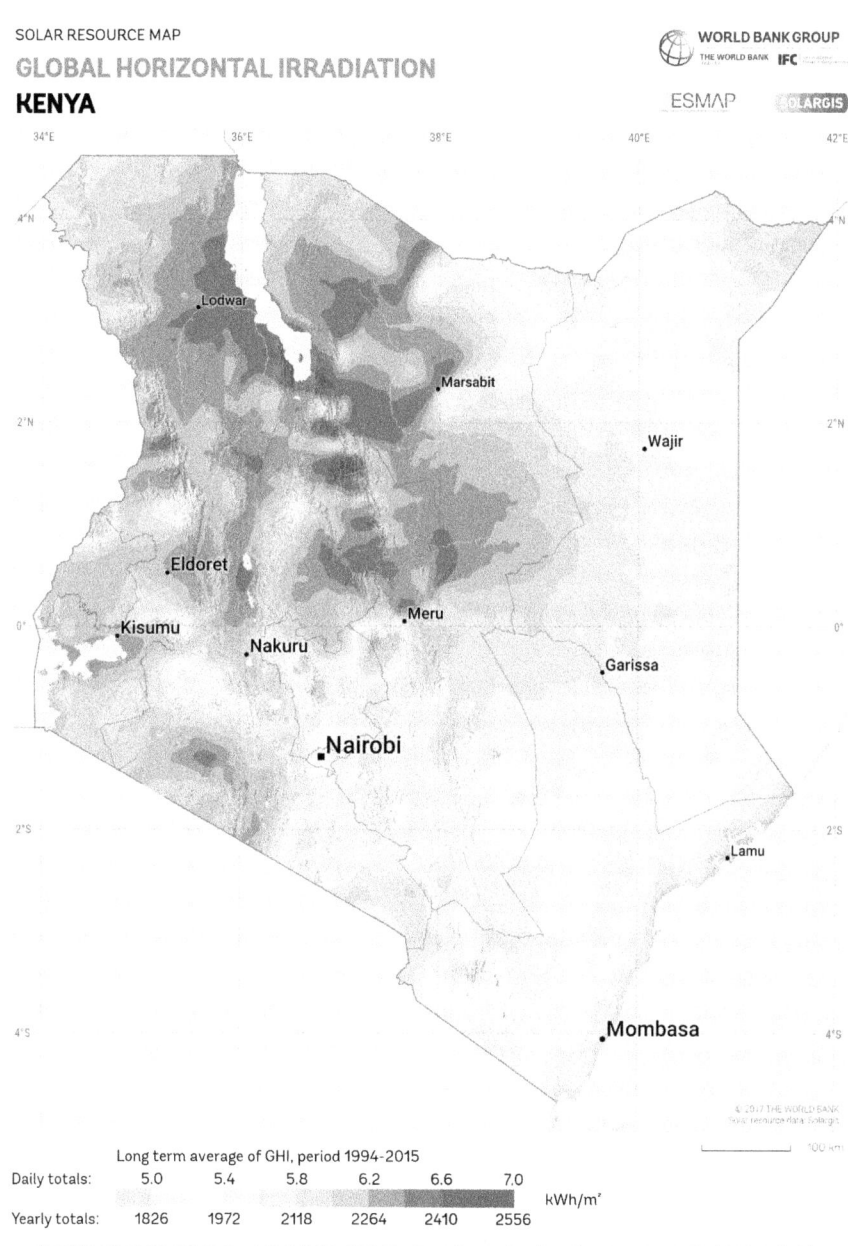

Figure B8 Solar resource map for Kenya
Source: Solargis

Step 5: Compute the system losses/performance ratio

Considering only the following losses:

Losses due to temperature, L_{temp}	= 10 per cent
Wiring losses, L_{wire}	= 3 per cent
Losses due to soiling/dirt, $L_{soiling}$	= 5 per cent
Losses due to reflectance L_{ref}	= 3 per cent
Losses due to incorrect orientation, $L_{orientation}$	= 2 per cent
Losses due to incorrect tilt, L_{tilt}	= 3 per cent
Loss due to power tolerance, $L_{tolerance}$	= 3 per cent
Losses due to mismatching, $L_{mismatch}$	= 2 per cent
Losses due to conversion	= 3 per cent
Losses due to light-induced degradation, LLID	= 3 per cent

Then PR = $(1 - 0.1) \times (1 - 0.03) \times (1 - 0.05) \times (1 - 0.03) \times (1 - 0.02) \times (1 - 0.03)$
    $\times (1 - 0.05) \times (1 - 0.02) \times (1 - 0.03) \times (1 - 0.03)$
    = $0.90 \times 0.97 \times 0.95 \times 0.97 \times 0.98 \times 0.97 \times 0.97 \times 0.98 \times 0.97 \times 0.97$
    = **0.684**

Step 6: Calculate the PV peak power to be installed

From formula (vi)

$$P_{Peak} = \frac{P_1 \times \text{Hours of operation}}{PSH \times PR}$$

$$P_{Peak} = \frac{4.828 \times 7}{6 \times 0.684} = \frac{33.796}{4.104} = \mathbf{8.23 \text{ kW}}$$

Step 7: Select a suitable inverter/controller

Using electrical data for Grundfos inverters, shown below and matching to the Grundfos SP9-21, 4 kW, 9.75-9.6-9.8A, 3 × 400 V (see motor electrical data below), the appropriate inverter is the 5.5 kW inverter (see under controller data below) with rated output current of 12 A 3 × 380–415 V. The system voltage is V_{min_mppt} = 400 VDC (recommended V_{min_mppt} = 530–61 5 V), V_{max_input} system = 800 VDC.

Note the 4 kW inverter with a rated output of 9.6A has not been selected as it is right on the edge of the required motor current.

Grundfos otor Electrical data:	
Motor type:	MS4000
Rated power - P2:	4 kW
Power (P2) required by pump:	4 kW
Mains frequency:	50 Hz
Rated voltage:	3 × 380-400-415 V
Rated current:	9.75-9.60-9.80 A

MANUAL CALCULATION OF SOLAR SYSTEM 207

Grundfos otor Electrical data:	
Starting current:	460-500-530 per cent
Cos phi - power factor:	0.85-0.80-0.77
Rated speed:	2850-2865-2875 rpm
Start. method:	direct-on-line
Enclosure class (IEC 34-5):	IP68
Insulation class (IEC 85):	F
Built-in temp. transmitter:	no
Motor No:	79194510

	Grundfos controller data			
Voltage			3 × 208-240 V	3 × 380-440 V
Installation environment	Min. ambient temperature	[°C (°F)]	−10 (14)	−10 (14)
	Max. ambient temperature	[°C (°F)]	60 (140)	60 (140)
	Max. relative humidity	[%]	100	100
Electrical data	DC minimum MPP voltage	[VDC]	230	400
	DC recommended MPP voltage	[VDC]	290–336	530–615
	DC maximum input voltage	[VDC]	400	800
	AC input voltage	[VAC]	208–240	380–460
	AC rated output voltage	[VAC]	220	380–440
	Min. frequency	[Hz]	5	5
	Max. frequency	[Hz]	60	60
	Phases		3	3
	Enclosure class		IP66	IP66

		Grundfos high voltage range		
Power [kW (hp)]	Product number	Electrical data		
		Max. P2 [kW (hp)]	Rated output current [A]	Frame size
2.2 (3)	99044348	2.2 (3)	5.6	A
3.0 (4)	99044349	3.0 (4)	8	A
4.0 (5)	99044350	4.0 (5)	9.6	A
5.5 (7.5)	99044351	5.5 (7.5)	12	A
7.5 (10)	99044352	7.5 (10)	16	B
11 (15)	99044363	11 (15)	23	B
15 (20)	99044364	15 (20)	31	B
18.5 (25)	99044365	18.5 (25)	38	C
22 (30)	99044366	22 (30)	46	C
30 (40)	99044367	30 (40)	61	C
37 (50)	99044368	37 (50)	72	C

Step 8: Determine the quantity of modules and module configuration

Using a multicrystalline 270 W module (Trinasolar) with electrical data shown below, V_{mp} = 30.9 V, V_{oc} = 37.9 V

Electrical data (STC)					
Peak Power Watts-PMAX (Wp)*	265	270	275	280	285
Power Output Tolerance-PMAX (W)			0 ~ +5		
Maximum Power Voltage-VMPP (V)	30.8	30.9	31.1	31.4	31.6
Maximum Power Current-IMPP (A)	8.61	8.73	8.84	8.92	9.02
Open Circuit Voltage-VOC (V)	37.7	37.9	38.1	38.2	38.3
Short Circuit Current-ISC (A)	9.15	9.22	9.32	9.40	9.49
Module E_ciency m (%)	16.2	16.5	16.8	17.1	17.4

Temperature ratings	
NOCT(Nominal Operating Cell Temperature)	44°C (±2°C)
Temperature Coe_cient of PMAX	−0.41%/°C
Temperature Coe_cient of VOC	−0.32%/°C
Temperature Coe_cient of ISC	0.05%/°C

$$\text{Number of modules, P} = \frac{P_{Peak}}{\text{Module rating}} = \frac{8.23 \times 1000 \, (W)}{270 \, (W)}$$
$$= 30.4 \text{ minimum number of modules}$$

Using formula ix the number of modules in series is calculated using the two limits:

$$\frac{V_{min\_mppt} \, (\text{system})}{V_{mp} \, (\text{panel})} = 530/30.9 = 17 \quad \frac{V_{max\_input} \, (\text{system})}{V_{oc} \, (\text{panel})} = 800/37.9 = 21$$

The number of modules per string is a value between 17 and 21, but the number should be as high as possible while keeping the number optimal. The number of modules in all strings must also be equal.

Since P = M × N, two parallel strings of between 17 and 21 modules will give a PV generator of between **9.18 kW and 11.34 kW**.

Check: Using Grundfos sizing software (a step-by-step process for using the Grundfos sizing tool can be found at GLOSWI, 2018f), the sizing gives the results shown in Figure B2. The proposed configuration is 18 panels in series in two strings (**9.72 kW**)

Alternatively, using a Lorentz inverter with the electrical data yields shown:

Lorentz controller data	
Power max.	8.0 kW
Input voltage max.	850 V
Optimum Vmp**	> 575 V
Motor current max.	13 A
Efficiency max.	98 per cent
Ambient temp.	−30 ... 50 °C
Enclosure class	IP54

$$\frac{V_{min\_mppt}\,(\text{system})}{V_{mp}\,(\text{panel})} = 575/30.9 = 18.6 \qquad \frac{V_{max\_input}\,(\text{system})}{V_{oc}\,(\text{panel})} = 850/37.9 = 22.4$$

Two parallel strings of between 19 and 22 modules will give a PV generator of between **10.26 kW and 11.88 kW**.

Check: Using the Lorentz Compass software (a step-by-step process for using the Lorentz Compass sizing tool can also be found at GLOWSI, 2018f) the sizing results are shown in Figure B3. The proposed configuration is 20 panels in series in two strings (**10.8 kW**)

In both cases, the check done using sizing software give designs that produce more water than the required daily amount of 60 m³/day. Using computer-based software takes into consideration the hydraulic and electric load, the available solar resource, the power losses and the module characteristics to give a more precise, predictable and optimal design than the manual process discussed here.

Step 9: Determine if alternative power source is required

In the case where the water demand is higher than what solar alone can provide, an alternative source of power will be required to prolong pumping. In this example, extended pumping is not necessary as solar alone is sufficient to meet the demand of 60 m³/day.

For illustration, if the water demand in this case was, say, 100 m³/day, considering the worst month, it would be necessary to prolong pumping for 4–5 hours with either a generator or grid power (if present) to provide the deficit of 40 m³/day. The additional pumping would be done in the early morning, late evening, or at night when solar pumping cannot be achieved.

Estimation of daily pumped flow

The amount of water pumped is dependent on the available solar resource, day by day and month by month, as well as the pump power. Estimation of pumped water can be done by applying previous formulas and working in reverse.

$$P_1 = P_2 \div \eta_m \quad \text{and} \quad P_2 = P_h \div \eta_p \quad \text{Hence;} \quad P_1 = P_h \div (\eta_p \times \eta_m)$$

Also,

$$P_{Peak} = \frac{P_1 \times \text{Hours of operation}}{PSH \times PR} = \frac{P_h \times \text{Hours of operation}}{\eta_p \times \eta_m \times PSH \times PR}$$

$$= \frac{Q \times \rho \times g \times H \times \text{Hours of operation}}{\eta_p \times \eta_m \times PSH \times PR \times 3.6 \times 10^6}$$

It is known that Daily flow = Q × Hours of operation, and using ρ = 1000 kg/m³, g = 9.81 m/s²

$$P_{Peak} = \frac{\text{Daily flow} \times \rho \times g \times H}{\eta_p \times \eta_m \times PSH \times PR \times 3.6 \times 10^6} = \frac{\text{Daily flow} \times H \times 2.725^{-03}}{\eta_p \times \eta_m \times PSH \times PR}$$

Thus, daily flow can be estimated as: Daily flow = $\dfrac{P_{peak} \times \eta_{(p,m)} \times PSH \times PR}{H \times 2.725 \times 10^{-3}}$

Daily flow (m³/day) = P_{peak} (Wp) × $\eta_{(p,m)}$ × PSH × PR ÷ H(m) × 2.725

Quick reference of combined pump and motor efficiency ($\eta_{p,m}$) is between 25 per cent and 60 per cent. All other factors in the formulas are known from previous workings. Therefore, by knowing the different PSH values month by month, one can work backwards to estimate the daily flow for each month or season for the installed SPWS. Likewise, by estimating the required duty point, the required PV generator can also be calculated by rearranging the formulas.

ANNEX C
Example of calculation of losses due to non-optimum tilt angle of PV modules

Question: How can more energy be obtained with 1 kW_{pk} of PV modules mounted in Valencia (Spain)?

1. With the modules mounted horizontally on the roof.
2. With the modules mounted in the south face of the same building.

Use the PVGIS calculation tools available from the European Commission (2019) to solve the question.

Remember that in PV systems the azimuth orientation is defined as the deviation from south (for the northern hemisphere).

Solution: With the default values given by the program for the location 'Valencia, Spain', and selecting the option 'Optimize slope and azimuth', the value of $PSH_{optimum}$ = 2070 PSH (yearly in-plane irradiation in kWh/m²) can be obtained:

Summary

Provided inputs:	
Location [Lat/Lon]:	39.470, −0.376
Horizon:	Calculated
Database used:	PVGIS-CMSAF
PV technology:	Crystalline silicon
PV installed) [kWp]:	1
System loss [%]:	14

Simulation inputs:	
Slope angle [°]:	36 (opt)
Azimuth angle [°]:	−1 (opt)
Yearly PV energy production [kWh]:	1610
Yearly in-plane irradiation [kWh/m²]:	2070
Year to year variability [kWh]:	43.10
Changes in output due to:	
Angle of incidence [%]:	−2.5
Spectral effects [%]:	0.6
Temperature and low irradiance [%]:	−7.8
Total loss [%]:	−22.2

212 SOLAR PUMPING FOR WATER SUPPLY

The PV modules mounted horizontally on the roof are 0° tilted and the orientation doesn't matter. The modules mounted on the south face have a tilt angle equal to 90°. Using the program for these two cases the following results are obtained:

Simulation inputs:	
Slope angle [°]:	0
Azimuth angle [°]:	0
Yearly PV energy production [kWh]:	1360
Yearly in-plane irradiation [kWh/m²]:	1770
Year to year variability [kWh]:	27.00
Changes in output due to:	
Angle of incidence [%]:	−3.6
Spectral effects [%]:	0.4
Temperature and low irradiance [%]:	−7.7
Total loss [%]:	−23.2

Simulation inputs:	
Slope angle [°]:	90
Azimuth angle [°]:	0
Yearly PV energy production [kWh]:	1050
Yearly in-plane irradiation [kWh/m²]:	1370
Year to year variability [kWh]:	38.80
Changes in output due to:	
Angle of incidence [%]:	−5.4
Spectral effects [%]:	0.8
Temperature and low irradiance [%]:	−6.5
Total loss [%]:	−23.4

The irradiation for the horizontal modules is $PSH_{0°} = 1770$ and for the modules on the facade is $PSH_{90°} = 1370$. It is evident that the horizontal configuration produces more energy than with the modules mounted on the facade. The $L_{ori+tilt}$ factor for each case are the following:

$$L_{tilt0°} = \left(1 - \frac{1770}{2070}\right) = 0.145 \rightarrow 14.5\%$$

$$L_{tilt90°} = \left(1 - \frac{1330}{2070}\right) = 0.375 \rightarrow 35.7\%$$

EXAMPLE OF CALCULATION OF LOSSES DUE TO NON-OPTIMUM TILT ANGLE

In the parameters given by the PVGIS tool is detailed the 'system loss', which corresponds to the performance ratio defined previously. It can be seen that, with a small variation, the three values of PR only vary by 1.1 per cent, with an important variation of the **losses due to angular and spectral reflectance** in the different configurations.

ANNEX D
Cable sizing

For pumping systems, the right dimension of the cable cross-section can be found using any of the following four ways:

Method 1: Computer-based pumping design simulations will give recommended cable size

Method 2: Manual calculation of the allowable voltage drop, followed by selection of the right cable size from voltage drop charts such as the one in Figure D1. In this method, the voltage drop permissible is computed in millivolts per amp per metre (mV/A/m). Alternatively for DC cables, the allowable voltage drop can also be calculated using the formula

$$\text{Voltage drop} = L_c \times R \times I$$

Where L_c = Length of a two-core cable (back and forth) in metres, I = Nominal current in Amps, R = Electrical resistance of a two-core cable in ohms/metre.

Method 3: Using tables provided by pumping equipment providers. These tables will usually have a list of cable sizes with the maximum length of each cable size that can be run from the motor to the power source for each motor size, based on a maximum voltage drop (1–3 per cent). An example of a submersible cable sizing chart provided by Grundfos is shown in Figure D2 (for AC motors at maximum water temperature of 30°C). Different tables are available for DC motors and DC cables.

Method 4: Using the formulas provided in section 4.3. It should be noted that for DC cables, the current flows all the way through the negative and positive circuit and therefore voltage drop is considered for double the length of cable.

$$\Delta P_{wire} = \Delta V_{wire} I$$

$$\Delta P_{wire} = R_{wire} \cdot I^2 = \rho \cdot \frac{l_{wire}}{S_{wire}} \cdot I^2 = \frac{1}{\gamma} \cdot \frac{l_{wire}}{S_{wire}} \cdot I^2$$

Example 1: AC cable sizing using method 2

11 kW 3 × 400 VAC, 24.6 A motor, 200 m AC cable

Note that the entire cable length should be considered from the pump all the way to the AC controller e.g. 150 m submerged + 50 m on the surface = 200 m.

Step 1 – Permissible voltage drop calculation: 3 per cent × 400 V × 1000 = 12,000 mV

Step 2 – Computation of permissible voltage drop per amp per metre through the cable

= 12,000 mV ÷ (24.6 A × 200 m) = 2.44 mV/A/m

Conductor or size (mm²)	Multicore armoured pvc insulated cable (PVC-SWA)				Twin and multicore pvc insulated cable			
	Two core cable Single phase supply		Three or four core cable Three phase supply		Two core cable Single phase supply		Three or four core cable	
	Max current capacity (A)	Voltage drop per amp per meter (mv)	Max current capacity (A)	Voltage drop per amp per meter (mv)	Max current capacity (A)	Voltage drop per amp per meter (mv)	Max current capacity (A)	Voltage drop per amp per meter (mv)
1.5	22	29.00	19	25.00	19.5	29.00	17.5	25.0
2.5	31	18.00	26	15.00	27	18.00	24	15.0
4.0	41	11.00	35	9.50	36	11.00	32	9.5
6.0	53	7.30	45	6.40	46	7.300	41	6.4
10.0	72	4.40	62	3.80	63	4.40	57	3.8
16.0	97	2.80	83	2.40	85	2.80	76	2.4
25.0	128	1.75	110	1.50	112	1.75	96	1.5
35.0	157	1.25	135	1.10	138	1.25	119	1.1
50.0	190	0.94	163	0.81	168	0.94	144	0.81

Figure D1 Table of cable current capacity and voltage drop
Source: Davis & Shirtliff Ltd

CABLE SIZING

Cable dimensions at 3 x 400 V, 50 Hz, DOL
Voltage drop: 3 per cent

Motor	kW	I_n [A]	Cos φ 100 %	\multicolumn{14}{c}{Dimensions [mm²]}
				1.5	2.5	4	6	10	16	25	35	50	70	95	120	150	185	240	300
4"	0.37	1.4	0.64	462	767														
4"	0.55	2.2	0.64	294	488	777													
4"	0.75	2.3	0.72	250	416	662	987												
4"	1.1	3.4	0.72	169	281	448	668												
4"	1.5	4.2	0.75	132	219	348	520	857											
4"	2.2	5.5	0.82	92	153	244	364	602	951										
4"	3	7.85	0.77	69	114	182	271	447	705										
4"	4	9.6	0.8	54	90	143	214	353	557	853									
4"	5.5	13	0.81	39	66	104	156	258	407	624	855								
4"	7.5	18.8	0.78	28	47	75	112	185	291	445	609	841							
6"	4	9.2	0.82	55	91	146	218	359	566	867									
6"	5.5	13.6	0.77	40	66	105	157	258	407	622	850								
6"	7.5	17.6	0.8	29	49	78	117	193	304	465	637	882							
6"	9.2	21.8	0.83	23	39	62	93	154	243	372	510	706	950						
6"	11	24.8	0.83		34	53	80	132	209	320	440	610	823						
6"	13	30	0.81		28	45	68	112	176	270	370	513	690	893					
6"	15	34	0.82			39	59	97	154	236	324	449	604	783	947				
6"	18.5	42	0.81				48	80	126	193	265	366	493	638	770	914			
6"	22	48	0.84				41	67	107	164	225	313	422	549	665	793	927		
6"	26	57	0.84					57	90	138	189	263	355	462	560	667	781	937	
6"	30	66.5	0.83					49	78	119	164	227	307	398	482	574	670	803	926
6"	37	85.5	0.79						63	97	133	183	246	317	382	452	525	624	714
8"	22	48	0.84				41	67	107	164	225	313	422	549	665	793	927		
8"	26	56.5	0.85					57	90	138	189	263	356	464	563	672	787	947	
8"	30	64	0.85					50	79	122	167	233	314	409	497	593	695	836	968
8"	37	78.5	0.85						65	99	136	190	256	334	405	483	567	682	789
8"	45	96.5	0.82						54	83	114	158	213	276	334	396	462	553	636
8"	55	114	0.85							68	94	131	177	230	279	333	390	469	544
8"	63	132	0.83								83	115	155	201	243	289	338	404	466
8"	75	152	0.86								70	97	132	171	208	249	292	353	409
8"	92	186	0.86									79	107	140	170	204	239	288	335
8"	110	224	0.87										89	116	141	169	198	240	279
10"	75	156	0.84							69	96	130	169	205	244	285	343	396	
10"	92	194	0.82								79	106	137	166	197	230	275	316	
10"	110	228	0.84									89	116	140	167	195	234	271	
10"	132	270	0.84										98	118	141	165	198	229	
10"	147	315	0.81											103	122	142	169	194	
10"	170	365	0.81												105	122	146	168	
10"	190	425	0.79													106	125	144	
12"	147	305	0.83											105	125	146	175	202	
12"	170	345	0.85												92	110	129	155	180
12"	190	390	0.84													98	114	137	158
12"	220	445	0.85														100	120	139
12"	250	505	0.85															106	123
Max. current for cable [A]*				23	30	41	53	74	99	131	162	202	250	301	352	404	461	547	633

* At particularly favourable heat dissipation conditions. Maximum cable length in metres from motor starter to pump.
For motors with star-delta starting, the cable length can be calculated by multiplying the relevant cable length from the above table by $\sqrt{3}$

Figure D2 Table of cable sizes
Source: Grundfos

Step 3 – Selection of the cable size from appropriate cable chart. From Figure D1 (for multicore armoured PVC insulated cable >> three or four core cable three phase supply) a **16 mm²** cable will have a voltage drop of 2.4 mV/A/m which is within the permissible maximum of 2.44 mV/A/m computed above.

Example 2: AC cable sizing using method 3

11 kW 3 × 400 VAC 24.6 A motor, 200 m AC cable.

From Figure D2, 2.5 mm² connected to an 11 kW motor can be run up to 34 m, 4 mm² up to 53 m, 6 mm² up to 80 m, 10 mm² up to 132 m and 16 mm² up to 209 m. Therefore, **16 mm²** should be selected as it accommodates the distance of 200 m.

Example 3: DC cable sizing using method 2

649 VDC, 18.4 A, 200 m DC cable

Step 1 – Permissible voltage drop calculation: 3per cent × 649 V × 1000 = 19,470 mV

Step 2 – Computation of permissible voltage drop per amp per metre through the cable

$$= 19,470 \text{ mV} \div (18.4 \text{A} \times 200 \text{ m}) = 5.29 \text{ mV/A/m}$$

Step 3 – Selection of the cable size from appropriate cable chart. From Figure D1 (for twin and multicore PVC insulated cable >> two core cable single phase supply) a 10 mm² cable will have a voltage drop of 4.4 mV/A/m, which is within the permissible maximum of 5.29 mV/A/m computed above.

Note since the cable is two core, it will be stripped so that one core carries positive and the other core carries negative.

In general, the thicker the cable, the lower the voltage drop and consequently to reduce operating losses the cable cross-section can be increased if it is economical and the conditions are favourable to do so.

Example 4: DC cable sizing using method 4

649 VDC, 18.4 A, 100 m DC cable (for DC double length will be considered i.e. 200 m)

Step 1 – Permissible voltage drop calculation: 3 per cent × 649 V = 19.47 V

$$\Delta P_{wire} = 19.47\text{V} \times 18.4\text{A} = 358.248\,\text{W}$$

$$S_{wire} = \frac{\rho \times L_{wire} \times I^2}{\Delta P_{wire}} \text{ for cable at 70°C, the resistivity}, \rho = 0.02136$$

$$= (0.02136 \times 200 \times 18.4^2) \div 358.248 = 4.04 \text{ mm}^2 \times 2 \text{ cores}$$

$$= 8.08 \text{ mm}^2 \text{ nearest cable size is 10 mm}^2$$

ANNEX E
Product warranty card sample

Format of warranty card to be supplied with each solar water pumping scheme

Item description
Solar PV modules
Make
Type of cell
Date of installation of modules
No. of modules installed
Serial no.'s
Rating of each PV module
Output voltage (V_{mpp}) of each module
Max output voltage of PV array
Warranty validity date
Pump-motor assembly
Make
Model/Part number
Serial number
Rated hydraulic capacity
Manufacturer production date
Installation date
Commissioning date
Warranty validity date
DC-AC inverter/pump controller/drive
Make
Model/Part number
Rating
Serial number
Installation date
Commissioning date
Warranty validity date
Isolation, earthing, and surge protection
Commissioning date
Warranty validity date
Designation and address of firm and contact person for claiming of warranty obligations

Plate and date   (Signature)
                          Name and designation
                 Name & address of manufacturer/supplier
                            (SEAL/STAMP)

ANNEX F
Routine inspection and maintenance activity sheets

Weekly activity sheet

	Regular inspection and preventative care activity sheet	
System component	Preventative care and maintenance	Activity
Pump (over and above regular operations procedures provided by operator)	Check area around pump for rubbish and debris	Remove and dispose of rubbish and debris
	Is pump or discharge piping leaking?	If yes, request technical support to fix leaks
	Record pump pressure when running	If there is a change from normal range in pressure, request technical support
Pumphouse/ Pump enclosure	Check enclosure for cracks and damage	If cracked, repair cracks with cement
	Is the enclosure locked?	If not, lock enclosure and/or repair locking hardware as necessary
Controllers/ Inverters, etc.	Record electrical discharge	If there is a change in electrical discharge, request technical support to check panels
Solar photovoltaic array	Check for a source of shade on the panels, such as vegetation or structures	Trim or remove any vegetation around the solar panels as well as any structures that will block sunlight
	Check area around solar panels for rubbish, debris, and spider webs or any other insect nesting	Remove and dispose of rubbish and debris. Carefully remove webs and nests
	Wash panels	Do during early hours when it is not yet hot. Use a soft sponge and water only
	Are there any cracks in the panels?	If yes, request technical support
	Is there any exposed or loose or disconnected wiring? Check for any damage from rodents or animals	If yes, request technical support
	Is the panel mounting strong and well attached? Are there cracks or any other signs of weakening?	If yes, request technical support

Regular inspection and preventative care activity sheet

System component	Preventative care and maintenance	Activity
Pressure line	Are any pipes exposed?	If yes, bury pipe
	Are any valves leaking?	If yes, request technical support
	Are any pipes leaking?	If yes, request technical support
	Are any connections leaking?	If yes, request technical support
	Are any valves stuck (cannot be moved), loose, or missing?	If yes, request technical support
Tank	Has the tank lid been damaged or removed?	If the lid is missing, replace the lid. If the lid is damaged, request technical support to fix the lid or place a purchase order for a new lid
	Is the tank (itself) cracked or leaking?	If yes, request technical support
	Are connections to the tank leaking?	If yes, request technical support
	Is the tank stand (support) damaged in any way?	If yes, request technical support
	Is the tank leaning?	If yes, request technical support
Distribution line	Are any pipes exposed?	If yes, bury pipe
	Are any valves leaking?	If the valve is loose, tighten the fittings. If the fittings are still leaking, remove valve and replace gasket. Replace entire valve if necessary
	Are any pipes/connections leaking?	If the pipe is leaking at a fitting, tighten the fitting. If the leak continues, request technical support
	Are any valves stuck (cannot be moved), loose, or missing?	If yes, request technical support
Taps and standpipes	Are any taps loose?	If yes, tighten tap
	Are any taps leaking?	If yes, tighten tap. If it continues to leak, replace gasket or entire tap as necessary
	Are any taps stuck or blocked?	If yes, replace tap
	Are any tap handles missing?	If yes, replace handle/ tap
	Are any standpipes leaning or falling over?	Straighten and construct support for tap stand or secure to existing

Daily activity sheet

	Regular inspection and preventative care activity sheet	
System component	Preventative care and maintenance	Activity
General	Follow the official pumping schedule of the scheme	As indicated by scheme management
Pump (over and above regular operation procedures provided by manufacturer's operating manual)	Maintain a clean facility at all times	Remove and dispose of rubbish and debris, sweep area
	Record pump pressure when running	Pressure_____Date_____Time_____
		Pressure_____Date_____Time_____
		Pressure_____Date_____Time_____
	Record pumping times	
Bulk water meter	Record readings	• If bulk water meter is not working read power consumption instead (daily) • Report functionality problems immediately • Request for calibration if there are doubts about the accuracy of the readings or if the meter has not been calibrated for more than five years
	Read pressure gauge	
Controllers/ inverters, etc.	Record electrical power	
	Check any warning lights or alarms	• Low water level in the well • Intrusion • Power outage • Pump failure • Low pressure
	Read volt and ampere meters	
Solar array	Check solar array for any immediate needs	• Leaves or sticks that have been blown or fallen onto the panels • Bird droppings or major dirt build-up • Breakages from storms or other means • Wiring

ANNEX G
Preventive maintenance plan

Adapted from SolarPower Europe <www.solarpowereurope.org>

The abbreviations describe the importance and frequency of the maintenance tasks related to each component of the solar plant:

Q: quarterly
SA: semi-annual
Y: yearly
nYr: every n years.

Equipment	Task	Importance	Frequency
Modules	Integrity inspection and replacement	Minimum requirement	Y
	Check cleanliness of modules	Minimum requirement	Y
	Electrical measurement inspections	Minimum requirement	Y
	Thermography inspection	Recommendation	Y
	Checking clamps/bolts panel structure	Minimum requirement	Y
	Internal inspection of junction boxes	Recommendation	Y
Electrical boards and switches	Integrity check and cleaning	Minimum requirement	SA or Y
	Check labelling and identification	Minimum requirement	Y
	Electrical protections (including fuses, surge and others) visual and functional tests	Minimum requirement	Y
	Checking integrity of cables and state of terminals	Minimum requirement	SA or Y
	Measurement inspection	Best practice	Y
	Thermography inspection	Recommendation	Y
	Check cable tightening	Minimum requirement	Y
	Monitoring operation test	Best practice	Y
Cables (DC and AC)	Integrity inspection	Minimum requirement	SA or Y
	Check labelling and identification	Minimum requirement	Y
	Check cable terminals	Minimum requirement	Y
	Measurement inspection	Best practice	Y

Equipment	Task	Importance	Frequency
Control box/ Inverter	Integrity check and cleaning	Minimum requirement	SA or Y
	Document inspection	Best practice	Y
	Check labelling and identification	Minimum requirement	Y
	Check correct operation (on/off) by operator	Minimum requirement	SA or Y
	Check fuses and surge protections	Minimum requirement	Y
	Thermographic inspection	Best practice	Y
	Sensor functional verification	Minimum requirement	Y
Generator (for emergency back-up or in hybrid systems)	Integrity check and cleaning	According to manufacturer recommendations	
	General maintenance		
	Check correct operation		
	Replacement of filters		
Lights and electrical sockets	Integrity check and cleaning	Minimum requirement	Y
	Check correct operation	Best practice	Y
	Check conformity to local security standards	Best practice	Y
Water supply system	Integrity inspection of pipeline and reservoir (if any)	Minimum requirement	SA or Y
	Check water meter correct operation	Minimum requirement	SA or Y
	Check water readings are properly taken by operator	Best practice	SA or Y
Lighting protection (if applicable)	Integrity inspection	Minimum requirement	Y
Fences and gates	Integrity inspection	Minimum requirement	Y
Vegetation	Vegetation clearing	According to local conditions	Q, SA or Y
Drainage system	General cleaning	According to local conditions	Q, SA or Y
Buildings	Integrity check and cleaning	According to local requirements	
	Documentation inspection	Best practice	Y
	Check earthing	Minimum requirement	3 Yr
PV support structure	Integrity inspection	Minimum requirement	Y
	Check tightening	Minimum requirement	Y
Irradiation sensors	Integrity check and cleaning	According to manufacturer specifications and local conditions	Q
	Calibration		2 Yr
	Monitoring operational test		Y

PREVENTIVE MAINTENANCE PLAN

Equipment	Task	Importance	Frequency
Communication board/Remote monitoring	Functional communication check	Minimum requirement	Q
Intrusion detection system	Integrity check and cleaning	According to manufacturer specifications	Y
	Functional verification of intrusion detection		Y
	Functional verification of alarms/cameras		Q
	Specific maintenance		Y
Stock of spare parts	Inventory of stock	Minimum requirement	Y
	Visual inspection of stock conditions	Minimum requirement	Y
	Stock replenishment	Minimum requirement	Q

ANNEX H
General troubleshooting for SPWSs

Troubleshooting of solar-powered water systems is unique to the system components, which is dependent on the product brand and type. This table is only a quick guide for common problems that may be encountered with a solar pumping system and therefore reference should be made to the manufacturer manuals for a complete troubleshooting procedure.

Fault	Possible cause	Check	Remedy
The pump does not start/run	Power supply to the pump has been cut off by the tank-full switch or dry-run protection	Check if the tank-full light is red	Tank is full – pump will come on with water use
		Check if the source-low light is red	Pump will come on when source recharges
	No or low power coming from the PV generator	Check if the solar modules are clean	Clean the modules
		Check for failed connections	Rectify insecure/unsealed wires, loose cable connections, failed cable joints
		Check if wiring (series/parallel) is correct	Correct wrong wiring
	No or low power reaching the controller	Check if accessories are correctly wired	Rectify wiring
		Check for failed connections	Rectify insecure/unsealed wires, loose cable connections, failed cable joints
	The motor is not getting power on all three phases	Check for loose connections and correct wiring	Rectify loose connections and wiring
	The dry-run protection is faulty	Check if the dry-run sensor is working correctly	Rectify dry-run sensor
The pump light comes ON and OFF	Solar power is not strong enough for the pump to run	Check if the panels are clean	Clean the modules
		Check if there is a passing shadow	This does not indicate a problem. The pump should restart after normal delay
		Check if the weather is overcast	Wait for the weather to improve

Fault	Possible cause	Check	Remedy
Pump ON light shows red instead of green due to system overload	Excessive pressure	Check if there is a restriction in the pipe such as a blocked pipe, closed valve	Remove restriction, open valve
	Insufficient motor cooling	Check for high running current	Confirm pump position and match to guidelines in section 6.2.2
	Blocked motor or pump	Check for high running current	Switch off the pump and call the system service provider/ technician to rectify
	High water temperature	Check for high running current	Switch off the pump and call the system service provider/ technician to rectify
Pump attempts to start at intervals (will start to turn or just vibrate a little)	Insufficient power reaching the controller	Check if accessories are correctly wired	Rectify wiring
		Check for failed connections	Rectify insecure/ unsealed wires, loose cable connections, failed cable joints
	Pump is running in reverse direction	Check if phases are correctly connected	Rectify wiring
	Motor is not getting power on all three phases	Check for loose connections	Tighten connections
	Pump or pipe is blocked	Check if pipes or pump are packed with mud, sand, debris	Clean pump and pipes
Pump runs but no water delivery	Surface valves are shut	Check if any valves are closed	Open all valves
	Broken downpipe	Check pipes by listening near borehole	Pull out the pump and replace worn out pipes
	Insufficient irradiation to deliver water	Check if irradiation threshold is enough for water delivery	Wait for the weather to improve
	Pump running in reverse direction	Check if phases are correctly connected	Rectify wiring
	The non-return valve is stuck in closed position	Check non-return valve	Clean or replace the valve
	The suction strainer is blocked		Pull out the pump and clean the strainer
	Pump/motor coupling broken	Check for low running current	Switch off the pump and call the system service provider/ technician

GENERAL TROUBLESHOOTING FOR SPWSs

Fault	Possible cause	Check	Remedy
Pump runs but there is lower water delivery than expected	Leaking downpipe or delivery pipe	Check pipes by listening near borehole	Replace worn-out pipes and fittings
		Check surface pipes for leakages	
	Source water level too low	Check if flow is intermittent and if motor current is low	Wait for source to recharge
	Pump or pipe chocked with silt/mud, valves partially closed	Check if water output is discoloured and silt laden	Unclog pipes, pull out pump and unclog
		Check if any valves are closed	Open valves
	The valves in the discharge pipe are partly closed/blocked		Clean or replace the valves
	Pump running in reverse direction	Check if phases are correctly connected	Rectify wiring
	Insufficient power reaching the controller	Check if DC accessories are correctly wired	Rectify wiring
		Check for failed connections	Rectify insecure/ unsealed wires, loose cable connections, failed cable joints
	Pump internal parts worn out		Call service centre for repair or replacement
Pump starts and stops at frequent intervals	The water level sensors or level switches in the reservoir are installed incorrectly	Check if sensor and switch intervals are too close	Adjust sensor/switch intervals

ANNEX I
Financing instruments for solar-powered irrigation systems

Commercial bank loan

The commercial bank loan is the most common instrument. The borrower requests funds from a finance intermediary (the bank) and they both agree on a repayment schedule and terms (interest rate). The intermediary has the choice to finance different projects in all sectors of the economy. The objective of the bank is to have the lowest default risk for the highest return. Intermediaries would usually try to build their portfolio by mixing high return (new technologies, for instance) with low risk (real estate) investments. Agriculture is seen as a risky sector as it depends on external parameters (climate variability). Unforeseen climatic events can drastically reduce the revenues of farmers (crop failures). Chapter 8 demonstrated that irrigation can reduce rainfall uncertainty. Bankers will ask for a lot of personal data to assess the credit worthiness of potential clients: age, sex, marital status, and bank and credit history. Finally, the intermediary will ask for collateral, which usually takes the form of land titles, which many farmers do not possess. The risk assessment of banks leads to different interest rates: 15 per cent in Senegal and 20–30 per cent in Kenya, for example (FAO & GIZ, 2018a).

Rural or development bank loans

These are an alternative to commercial bank loans. The portfolio of these banks is composed of only rural development projects in which agriculture is the major share. As such, they operate in a riskier environment. They can offer lower interest rates due to their knowledge of the sector but they have a lower earning capacity and are subject to competition from minor and more flexible institutions, such as microfinance actors. For this reason, they would usually partially finance the investment (around 75 per cent), meaning that farmers need to possess seed capital. The interest rates of agricultural banks are lower than commercial banks. For instance, the interest rate charged by the Agricultural Bank of Ghana for personal loans is the base rate plus 4–8.5 per cent; for the agriculture and forestry sector, it is the base rate plus 0–5 per cent (ADB, 2019). Finally, famers still need a bank account and collateral. At this stage, it seems that neither commercial nor agricultural banks are encouraged to finance SPISs: awareness is still lacking on the specificities of the equipment and their advantage (in case of default, solar systems could be used for other uses). International funding agencies could refinance local banks with specific lines for clean energy to encourage local banks to finance SPISs.

Microfinance

Microfinance institutions (MFIs) manage risk on their loan by reducing the size of the investment and imposing frequent repayments (monthly, weekly or even daily) with short contract durations. Typical loans from MFIs range between US$100 and $300, which may not be sufficient for systems pumping groundwater. However, this amount could complement farmers' own investments to cover the purchase of a piston pump to lift water from a surface source. In 2008, the global average microcredit interest rate was 35 per cent (Kneiding & Rosenberg, 2008) well above commercial and rural development banks.

Value chain loans

Input providers (upstream) or traders (downstream) can provide value chain loans. Input providers will provide inputs at a reduced cost before harvesting. After harvest, the farmer will pay the intermediary the remaining costs of the inputs plus a fee. Traders will lend money to farmers before harvest to purchase inputs. After harvest farmers will receive a payment from traders for their production minus the costs of inputs and a fee. In the case of SPISs, pump providers could establish contracts with farmers for a delayed payment of part of the equipment cost after harvest. Downstream value chain loans (traders) are not specific to SPISs and might only be used if the SPIS has a shorter payback period than other irrigation technologies. For the end users, the interest rate can be as high as 30 per cent since financial intermediaries' interest rates will be summed up with the pump provider interest rate based on its risk analysis. For instance, Futurepump in Kenya provides pumps to farmers. They can contract a loan with a commercial bank at a reduced interest rate (14 per cent and two years maturity) with a deposit corresponding to 30 per cent of the pump cost and a 5 per cent set-up fee (FAO & GIZ, 2018a).

Leasing or repurchase agreements

These are financial instruments in which the farmer leases the pump and linked solar system. At the end of the lease period, the farmer has the choice to buy the equipment or not. The user pays the lessee two fees: a utilization fee and a depreciation fee. The lessor pays the price of the equipment plus an interest rate. The leasing fees are paid monthly. Leasing is cheaper for farmers compared to loans. In the event of non-payment the financial intermediary can claim the equipment back from the farmer.

Agrarian cooperatives

These can be financed by agricultural and commercial banks, and by governmental and international development programs. They are legally registered entities with a specific status in most countries. They are also financed by contributions from their members. Thanks to their membership base and their legal status, they usually appear to banks to be less risky than individual farmers. Therefore, they are able to offer loans to their members for equipment, including SPISs, with reasonable interest rates. In cooperatives, default is less important than with commercial banking because cooperative members help each other to reimburse the loan taken collectively by an individual farmer (Huppi & Feder, 1990). Furthermore, in case of default, another member can repurchase the equipment.

Informal saving groups

This category covers different types of organization and schemes, such as village saving loan associations (VSLAa), tontines, table finance or informal loans. The aim is to gather savings from individual group members to finance investment for the whole group or for one of the members of the group. For instance, in a VSLA project in Rwanda, the loan amount during the first loan cycle ranged between US$200 and $1,000, which is similar to the amount offered by microfinance institutions (CARE, 2007). VSLA and agrarian cooperatives could support financial and operational models where wells are drilled in common, investments done by the group, and maintenance paid collectively, or for small surface pumps irrigating communal fields.

Pay-per-use business models

Pay-per-use comes in two types of investment mechanism: the fee-for-service model and the pay-as-you-go scheme (Moving Energy Initiative, 2018). In the fee-for-service model, farmers make an initial payment to cover the installation costs of the SPIS. The consumer makes regular payments to unlock the energy device. This could work either by time used, volume of water, or amount of energy. In the pay-as-you-go model, the consumer does not make the initial payment and makes regular payments to use the energy device and linked water volume for a fixed period of time or fixed volume. There are different ways to pay for the service: SMS or scratch top-up cards.

Subsidy mechanisms

Indirect subsidies exist on solar equipment when countries do not apply value-added tax (VAT) on solar products. Specific subsidy mechanisms for SPISs exist at national level in India, Nepal, and Tunisia (FAO, 2019). In India, subsidies depend on the different states. They range from 100 per cent of the total cost of equipment in Bihar to 60 per cent of the total cost in Haryana. They represent 70 per cent of the total cost for a woman farmer in Nepal. In Tunisia, the subsidy is 40 per cent of the total equipment cost with a US$7,000 limit for the subsidy per project. The subsidy obtained could be used as collateral to secure bank loans for SPISs. Subsidies can also be used as leverage to support the development of the SPIS sector together with water efficiency measures. For instance, different premiums could be offered to farmers depending on their project. A higher percentage of the project could be covered if the farmer implements drip irrigation or for collective use of the SPIS, for example.

Matching grants mechanisms

These use the same principle as subsidy mechanisms but with a different origin of fund: cooperation funds instead of national or federal state funds (IWMI, 2018).

ANNEX J
Physical control installation and maintenance checklists

In order to get a good-quality solar pumping installation, monitoring the field work carried out by the private contractor selected is of paramount importance. It is therefore strongly recommended to WASH officers to follow up as much as possible this list of actions.

Introduction: 4 steps

1. Check the references of all components of the system to ensure that the installed components are those provided in the design.
2. Check orientation and the inclination of the panels, and shadow on the solar PV generator. The orientation and inclination values must be close enough to those that were determined during calculation sizing. The acceptable variations will be less than 5° for the inclination and 15° with respect to the geographic north–south orientation.
3. Check the cleanliness and protection of the wiring, and its compliance with the standards.
4. Finally inspect civil works (e.g. castle, basin, trough, fixing the solar supports), piping, valves and all other important elements that can compromise the sound operation of the system.

Modules and PV array

No.	Subject	Observations
1.	Check the conformity of module specification in accordance with the design simulation and that all installed modules are of the same characteristics	
2.	Check the number of modules: number of modules in series and in parallel and compare with design	
3.	Check with the help of a compass if the modules are well oriented in south or north direction (according to the hemisphere), by positioning the compass against the east or west edge of a module	
4.	Check the inclination of the modules (tilt angle) using the inclinometer	
5.	Check with a level if the east–west axis of the modules is properly horizontal	
6.	Check that the height of the lowest point of the modules from the ground is greater than or equal to the height in the specifications	

No.	Subject	Observations
7.	Check the cleanliness of the solar array (each cell)	
8.	Ensure that no module is damaged, e.g. broken glass, frame, twisted, scratched	
9.	With the installer, measure the voltage output from each string and ensure it's consistent with the design and uniform across all the strings	
10.	Check and ensure that every module is fastened to the structure at every bolt hole in a manner that is robust, acceptable, and will deter vandalism	

Electrical wiring

No.	Subject	Observations
1.	Ensure cable conformity: compare cable specifications and sections with those provided by the manufacturer for the power and the distances measured on site	
2.	Check that all cable connections are inside the junction boxes provided for this purpose; no connection between two cables should be visible	
3.	Ensure all cable glands and conduits match the cable size and are properly sealed. All cable entries into the terminal box should be through cable glands	
4.	Check that all housings connections are at a minimum height of 50 cm from the ground	
5.	Choose a sample of the cable and test it by pulling a cable out of a gland to ensure that it is sufficiently tight to hold the cable	
6.	Check that all cable terminals are properly fastened and sufficiently tight	
7.	Check the cable interconnection between the modules is fastened to the structure at regular intervals by use of suitable clips or cable ties	
8.	Check that all surface cabling is of armoured type. If not armoured, it should be placed in electrical conduits and protected using protective tiles to prevent damage from passing vehicles	
9.	Check that there are no overhead cables. All interconnecting cables (e.g. connecting two support structures) should be guided to the ground and conform to point 7 above	
10.	Verify the existence and proper connection of grounding rods for both earthing and lightning surge protection	

Solar support structure

No.	Subject	Observations
1.	Check if all the members of the supporting structure are of the material specified. Ensure that no parts are susceptible to corrosion. Check for proper and uniform painting of the structure	
2.	Check the proper eye alignment of the support posts	

PHYSICAL CONTROL INSTALLATION AND MAINTENANCE CHECKLISTS

No.	Subject	Observations
3.	Check with a spirit level the verticality and horizontality of poles and modules (this allows to check the general quality of the completed work)	
4.	Check the bolting has been done at every hole	
5.	Check if the foundations are of sufficient size	
6.	Check for obvious weaknesses such as structures that are grossly swaying or leaning	

Inverter or other AC interface

Note: The pumps are equipped with an 'inverter box' also serving as a control box. For those who work in DC, the control cabinet is often called 'interface' or power conditioner.

No.	Subject	Observations
1.	Check the conformity of inverter specifications (or interface)	
2.	Check for varistors presence between positive terminal and earth; and between terminal negative and earth (or between positive and negative terminals if the terminal negative is connected to earth) on the terminal block to the input of the inverter (or interface)	
3.	Check the inverter (or interface) is correctly mounted at more than 50 cm above the ground	
4.	Make sure inverter is well protected from adverse weather conditions and is as close as possible to the PV array e.g. placed in the shade of the modules. If mounted inside a room, sufficient ventilation should be provided. The inverter must not be installed inside an additional enclosure as this will lead to insufficient cooling	
5.	Check that the inverter is mounted directly on a solid wall or equipped with a backplate, and that the wall/backplate can support the weight of the inverter. Check to ensure that it is also mounted in accordance with minimum spacing requirements provided by the manufacturer	
6.	Ensure protective devices have been installed between the PV array and the inverter e.g. DC disconnection switches, DC breakers, surge protectors	

Pump unit

No.	Subject	Observations
1.	Check the conformity of motor specifications	
2.	Check the conformity of pump specifications	
3.	Check that the splicing kit is of suitable type, quality, and workmanship. Resin type joint (not heatshrink) is recommended for deep installations. Check for obvious mistakes like air pockets on the joint, uniformity of mould	

No.	Subject	Observations
4.	Check that the pump setting depth complies with the purchase order/TOR	
5.	During installation, check that the drop cable and well probe cable are fastened to the pipe using suitable cable ties (tape or metallic clips MUST NOT be used) and at intervals of 3 m. Ensure the cable has been fastened with a stretching allowance as the delivery pipe is subject to elongation when filled with water (about 2 per cent stretch)	
6.	During the test run: Check that the protective control features are set in accordance with the pump specifications e.g. speed, voltage, sensor settings Check that the pump performance is in accordance with the design in terms of flow and pressure (e.g. reversed phases will cause reduced flow) Check that the current consumed by the pump is consistent with the pump specification	

Drill head

No.	Subject	Observations
1.	Ensure that all the components of the drill/wellhead are of corrosion-resistant material	
2.	Check that the gate valve meets the specifications and that it is fully open. Valve handle should be removed to avoid accidental closure	
3.	Check the conformity of the water meter to the specifications and mounting direction. Monitor compliance with the minimum safety lengths: 20 times the nominal diameter upstream, 10 times downstream	

System monitoring and control

No.	Subject	Observations
1.	System registered on remote monitoring platform	
2.	Controller registered/activated for data logging and remote access provided	
3.	Alerts and notifications activated to system owner's email/SMS	

General

No.	Subject	Observations
1.	2 days hands-on training of operators/users/NGO done on site	
2.	Testing, installation and commissioning report handed to NGO or asset owner if other.	

ANNEX K
Daily photovoltaic module and pump operation/monitoring format

Adapted from World Bank

Daily PV array and pump performance data

Recorded by: _____	Date_____	Designation _____
Location _____	District_____	Region _____
Latitude _____	Longitude_____	Elevation _____ (m)
Pump type _____	Pump duty point ____ (m^3/h)	Motor type _____
Motor rating (kW) _____	PV array type_____	PV cell type _____
PV array peak Watt _____	No. of panels _____ (°C)	Inverter make _____
Inverter type _____	Inverter rating _____ (kW)	Inverter rating _____(Amps)
Borehole depth _____ (m)	Static water level _____(m)	Dynamic water level_____ (m)
Pump setting _____ (m)	Reservoir capacity ____(m^3)	

Hr of the day	Weather (rainy, cloudy, sunny)	Power input (solar, generator, other)	Actual PV voltage (VDC)	Actual inverter voltage (VAC)	Air temp (°C)	Cell temp (°C)	Solar radiation (W/m^2)	Flow meter reading (m^3)	Pressure reading (Bar)	Actual PV current (Amps)	Actual inverter current (Amps)
0700											
0730											
0800											
0830											
0900											
0930											
1000											
1030											
1100											
1130											
1200											
1230											
1300											
1330											
1400											
1430											
1500											
1530											

Hr of the day	Weather (rainy, cloudy, sunny)	Power input (solar, generator, other)	Actual PV voltage (VDC)	Actual inverter voltage (VAC)	Air temp (°C)	Cell temp (°C)	Solar radiation (W/m²)	Flow meter reading (m³)	Pressure reading (Bar)	Actual PV current (Amps)	Actual inverter current (Amps)
1600											
1630											
1700											
1730											
1800											
1830											

Glossary

alternating current (AC) – electric current in which the direction of flow oscillates at frequent, regular intervals over time.

AM – air mass.

Ampere or Amps – a measure of electric current.

altitude – the angle between the horizon (a horizontal plane) and the sun, measured in degrees.

amorphous silicon (a-Si) – a thin-film PV silicon cell having no crystalline structure.

aquifer – a naturally occurring layer of water-bearing soil, rock, or sand.

asset owner – this can be the stakeholder who contributes to financing of construction (donor or UN agency, NGO, government, or others) or the community of users if equipment is handed over to them.

asset manager – aims at ensuring optimal functioning of the solar pumping scheme (or a portfolio of schemes) by supervising energy and water production, and O&M activities. Reports to asset owners.

azimuth – angle between true south and the point directly below the location of the sun, measured in degrees.

balance of system (BoS) – components of a PV system other than the PV array.

battery – two or more cells electrically connected for storing electrical energy.

battery bank – an energy storage capacity (ampere-hour).

borehole – a hole drilled through the ground to reach water. Borehole diameters vary depending on the required size of the system. Standard borehole sizes are 4 to 12 inches (100 to 300 mm).

camp – a context of displaced persons living in a confined location managed by a UN agency, such as UNHCR or IOM, and whose services ae provided by implementing agencies, such as NGOs.

capital cost – initial investment of a project.

cavitation – phenomenon in which low pressure causes bubbles to be formed at the suction inlet of the pump.

CdTe – a type of PV cell made from cadmium and telluride.

centrifugal pump – a pump that delivers water centrifugally using impellers by producing a pressure difference.

CIGS – a type of PV cell made from copper, indium, gallium, and di-selenide.

CIS – a type of PV cell made from copper, indium, and selenide.

contractor – entity that is in charge of installation of the solar pumping components, and/or operation and maintenance of the solar-powered pumping system.

conversion efficiency – the ratio of the electric energy produced by a PV to the energy from incident sunlight.

crystalline silicon (c-Si) – a type of PV cell made from a single crystal or polycrystalline slice of silicon.

current – the flow of electric charge in a conductor between two points having a difference in potential (voltage).

days of autonomy – the number of consecutive days a stand-alone system will meet a defined load without energy input.

design month – the month having the lowest renewable energy production to load ratio.

diffuse radiation – solar radiation scattered by the atmosphere.

direct radiation – solar radiation transmitted directly through the atmosphere.

direct current (DC) – electric current flowing in one direction.

disconnect – switch gear used to connect or disconnect components in a stand-alone system.

discount rate – rate at which the value of money changes relative to general inflation.

drawdown – the distance below the water table that the water level in a well falls to when steady state pumping is in progress.

duty point – the required flow and head of the operating pump.

efficiency – the ratio of output power to input power, expressed as a percentage.

E-generated – energy generated and available for consumption in the AC and DC loads.

electric circuit – a complete path followed by electrons from a power source to a load and back to source.

electric current – the magnitude of the flow of electrons.

flatplate – an arrangement of solar cells in which the cells are exposed directly to normal incident sunlight.

g – temperature coefficient of power of the PV module.

global solar radiation – the sum of diffuse and direct solar radiation incident on a surface.

grid – the network of transmission lines, distribution lines, and transformers used in central power systems.

HIT – heterojunction with intrinsic thin-film layer.

hydraulic equivalent load – the product of the daily amount of water produced by the pumping head.

hydraulic energy – the energy necessary to lift water.

IEC – International Electrotechnical Commission

incidence angle – angle that refers to the sun's radiation striking a surface. A normal angle of incidence refers to the sun striking a surface at a 90° (or perpendicular) angle.

irradiance – the instantaneous power per unit area received by a surface on the earth's surface (expressed in W/m^2).

insolation – a measure of the cumulative irradiance received on a specific area over a period of time, measured in Wh/m^2 or kWh/m^2.

inverter – a solid-state device that converts a DC input to an AC output.

kilowatt (kW) – one thousand watts.

kilowatt hour (kWh) – one thousand watt hours.

W/m^2 – watts per square metre.

kW/m^2 – kilowatt per square metre.

LID – light-induced degradation.

life-cycle cost – an estimate of the cost of owning and operating a system for the period of its useful life, usually expressed in terms of the present value of all costs incurred over the lifetime of the system. The sum of the capital cost and the present worth of the recurrent, salvage, and replacement costs.

load – the amount of electrical power being consumed at any given moment. Also, any device or appliance that is using power. In SPWSs the load is the pump.

load matching – the process of matching the load with the input power source to maximize the power transfer to the load.

load-matching factor – a non-dimensional factor defined by the ratio of energy acquired by the hydraulic load to the maximum power extracted from the power source in a one-day period. It can also be the ratio of the actual power output used for water pumping to the power source output capability.

maximum power point tracker (MPPT) – the impedance-matching electronics used to operate a PV array output at its maximum power.

module (panel) – a predetermined electrical configuration of solar cells laminated into a protected assembly.

monocrystalline silicon – a material formed from a single silicon crystal.

net present cost (NPV) – all project expenses converted into the current value of money.

NOCT – nominal operating temperature of a cell.

nominal voltage – a reference voltage used to describe batteries, modules, or systems (e.g. a 12-volt or 24-volt battery, module, or system).

orientation – placement according to the cardinal points (N, S, E, and W); azimuth is the measure in degrees from true south.

O&M – operations and maintenance.

payback period – the number of years (periods) required for the income (benefit) from a project to equalize its investment cost.

peak hour demand – the maximum amount of water required in an hour. In most cases, peak hour can refer to noon and/or evening water consumption.

peak efficiency – the highest output efficiency level that a solar panel or a solar inverter can achieve.

peak sun hours (PSH) – equivalent number of hours per day when solar irradiance averages 1,000 W/m^2.

peak watt (Wp) – the amount of power a PV device will produce during peak solar radiation periods when the cell faces directly towards the sun (value at STC).

performance ratio (PR) – the ratio between the generated energy and the theoretical energy that would be generated by the PV field if the modules converted the irradiation received into useful energy according to their rated peak power.

photovoltaic (PV) – phenomenon of generation of electricity from the sun's energy.

photovoltaic (PV) cell – a cell that generates electrical energy when incident solar radiation impinges on it.

photovoltaic (PV) system – an integrated system composed of a PV array, power conditioning, and other subsystems, such as the motor–pump.

polycrystalline silicon – silicon that has solidified at a rate such that many small crystals have formed.

positive displacement pump – a type of water pump that can lift water from a borehole by means of a cavity or cylinder of variable size. Also called a volumetric/helical rotor pump.

power conditioning – the electrical equipment used to convert power from a PV array into a form suitable to meet the power supply requirements of more traditional loads. It is a collective term for inverter, transformer, voltage regulator, meters, switches, and controls.

present worth – the value of future costs or benefits expressed in the current value of money (present-day money).

pumping head – the height of a water column that would produce the pressure that the pump experiences.

PV array – a mechanically and electrically integrated configuration of PV modules and support structure designed to form a DC power-producing unit.

PV field/PV plant/PV generator – the entire PV module arrangement in a solar installation.

remote site (location) – a site that is not located near a utility grid.

renewable energy (RE) – energy produced by non-fossil fuel or nuclear means. Includes energy produced from PV, wind turbines, hydroelectric, and biomass.

rising main – the pipe used to lift water from the borehole or surface pumping source.

rotor – the rotating central section of a motor or a pump.

solar collector – a horizontal or tilted surface on the earth's surface that captures the solar radiation.

solar controller – an electric device used for controlling and monitoring a solar pumping system.

solar thermal electric – a method of producing electricity from solar energy by concentrating sunlight on a working fluid that changes phase to drive a turbine generator.

SPWS – solar-powered water system or solar-powered water scheme.

stand-alone system – a system that operates independently of utility lines. It may draw supplementary power from the utility but is not capable of providing power to the utility.

static head – the height over which water must be pumped. Static head may vary due to seasonal changes in well recovery rates, fluctuations in groundwater level, etc.

stator – the outer stationary component of a motor or a progressive cavity pump.

standard test conditions (STC) – defined as irradiance of 1,000 W/m², solar spectrum of AM 1.5 and module temperature at 25 °C.

string – set of solar modules electrically connected in series.

switchgear – the switches and electrical controls in a solar-powered water system.

tilt angle – angle of inclination of a PV array as measured in degrees from the horizontal surface. Usually equal to the latitude of the PV array's location.

tilted factor – the ratio of the incidence solar radiation on a tilted PV array surface to the global solar radiation.

TPV_cell – PV cell temperature.

volt (V) – a unit of measurement of the force given to electrons in an electric circuit; electric potential (voltage).

water table (static water level) – the level below the ground at which the natural water level can be found.

watt (W) – measure of electric power. Watts = volts x amps.

watt hour (Wh) – a quantity of electrical energy when one watt is used/generated for one hour.

zenith angle – the incidence angle to a horizontal surface.

References

ADB, 2019

Brandt, M.J., Johnson, K.M., Elphinston, A.J. and Ratnayaka, D.D. (2017) 'Dosing pump', in Brandt et al., *Twort's Water Supply – 7th Edition,* Chapter 12, <https://www.sciencedirect.com/topics/engineering/dosing-pump>.

Bridge to India (2019) *'Managing India's PV module waste'*, Bridge to India, Gurugram, <https://bridgetoindia.com/backend/wp-content/uploads/2019/04/BRIDGE-TO-INDIA-Managing-Indias-Solar-PV-Waste-1.pdf>.

CARE (2007). *Linkages between CARE's VSLAs with Financial Institutions in Rwanda.* <https://mangotree.org/files/galleries/691_CARE_Rwanda_SLA_Linkage_to_Credit_Case_study_-_August0712.pdf>.

C. Mateo, M.A. Hernández-Fenollosa, Montero, S. Seguí-Chilet, Analysis of initial stabilization of cell efficiency in amorphous silicon photovoltaic modules under real outdoor conditions, Renew. Energy. 120 (2018) 114–125. doi:10.1016/j.renene.2017.12.054.

Davis & Shirtliff Ltd (2014) Cable splicing for submersible pump: <https://youtu.be/ox4oak9Bjm8>.

Delegation of the European Union to Pakistan (2018) *Revival of Balochistan Water Resources Programme,* <https://eeas.europa.eu/delegations/pakistan/52697/eu-delegation-provides-technical-assistance-effective-water-and-land-management-balochistan_en>.

Energypedia (2019) *Solar PV in hot climate zones*, <https://energypedia.info/wiki/Solar_PV_in_hot_climate_zones>.

Energypedia (2020) *Solar Pumping Toolkit – The Global Solar & Water Initiative,* <https://energypedia.info/wiki/Solar_Pumping_Toolkit_-_The_Global_Solar_%26_Water_Initiative>.

European Commission (2019) Interactive tools. *Photovoltaic Geographical Information System* <http://re.jrc.ec.europa.eu/pvg_tools/en/tools.html>

FAO (1998) *Crop Evapotranspiration – Guidelines for Computing Crop Water Requirements*, FAO Irrigation and Drainage Paper 56, Chapter 4, Food and Agriculture Organization of the United Nations, Rome, <http://www.fao.org/3/X0490E/x0490e08.htm>.

FAO (2014) *Walking the Nexus Talk: Assessing the Water-Energy-Food Nexus in the Context of the Sustainable Energy for All Initiative*, FAO, Rome, <http://www.fao.org/3/a-i3959e.pdf>.

FAO (2015) *Opportunities for Agri-Food Chains To Become Energy-Smart*, FAO, Rome, <http://www.fao.org/3/a-i5125e.pdf>.

FAO (2017) *'Gambian farmers adapt to climate change with new irrigation strategies'*, FAO, Rome, <http://www.fao.org/in-action/new-irrigation-strategies-gambia/en/>.

FAO (2018a) *The Benefits and Risks of Solar-Powered Irrigation – A Global Overview*. FAO, Rome, <http://www.fao.org/3/i9047en/I9047EN.pdf>.

FAO (2018b) *Costs and Benefits of Clean Energy Technologies in the Milk, Vegetable and Rice Value Chains*, FAO, Rome, <http://www.fao.org/3/i8017en/I8017EN.pdf>.

FAO & GIZ (2018a) *Module 10: Finance of the SPIS toolbox*. <https://energypedia.info/wiki/Toolbox_on_SPIS>.

FAO & GIZ (2018b) *Toolbox on Solar Powered Irrigation Systems*, Energypedia, <https://energypedia.info/wiki/Toolbox_on_SPIS>.

FAO & GIZ (2019) *Impact des systèmes de pompage et d'irrigation à énergie solaire en Tunisie*.

FAO & ICIMOD (2019) *Bangladesh Policy Brief Focus Areas*.

GLOSWI (2018a) *Field visit reports*, <https://energypedia.info/wiki/Solar_Pumping_Toolkit_-_Monitoring_and_Evaluation#GSWI_visit_reports_2016-2017>

GLOSWI (2018b)<https://www.youtube.com/watch?v=3H-8qfC68EI>

GLOSWI (2018c) *Installation Checklist*, <https://energypedia.info/wiki/File:Installation_Control_Checklist.pdf>.

GLOSWI (2018d) *Sample bidding document*, <https://energypedia.info/wiki/File:Bidding_Template.pdf>.

GLOSWI (2018e) *Solar Pumping emergency kits*, <https://washcluster.net/gwc-resources>.

GLOSWI (2018f) *The Solar Pumping Toolkit*. WASH Cluster. <https://washcluster.net/node/131>.

GLOSWI (2018g) *Resources*, <https://washcluster.net/gwc-resources>.

Grundfos (2012) *Grundfos catalogue SP A, SP range*, <https://net.grundfos.com/public/literature/filedata/Grundfosliterature-1098>.

Grundfos (2018) *Grundfos submersible pumps data booklet*, <http://net.grundfos.com/Appl/ccmsservices/public/literature/filedata/Grundfosliterature-1098.pdf>.

Grundfos (2020) *Grundfos Product Centre: Sizing software*, <https://product-selection.grundfos.com/front-page.html?custid=GMA&qcid=835427824>.

Hagenah, M. (2017) Information from GIZ. Programme Energies Durable (PED).

Huppi, M. and Feder, G. (1990) 'The role of groups and credit cooperatives in rural lending', *World Bank Research Observer*, 5(2), 187–204, <http://documents.worldbank.org/curated/en/774751468152395258/pdf/770010JRN0WBRO0Box0377291B00PUBLIC0.pdf>.

ICRC (2010) *Borehole drilling and rehabilitation under field conditions*, <https://www.icrc.org/en/doc/assets/files/other/icrc_002_0998.pdf>.

IEA PVPS (2018) *Annual Report, 2018*, International Energy Agency, <http://www.iea-pvps.org/index.php?id=6>.

International Energy Agency (2019) *International Energy Outlook 2019*, Renewable. <https://www.iea.org/reports/world-energy-outlook-2019/renewables#abstract>.

IRENA (2019) *Renewable Power Generation Costs in 2018*, International Renewable Energy Agency, Abu Dhabi. <https://www.irena.org/-/media/Files/IRENA/Agency/Publication/2019/May/IRENA_Renewable-Power-Generations-Costs-in-2018.pdf>.

IWMI (2018) *Business model scenarios and suitability: smallholder solar pump-based irrigation in Ethiopia*. Agricultural Water Management – Making a Business Case for Smallholders. International Water Management Institute (IWMI).

Kneiding, C. and Rosenberg, R. (2008) *'Variations in microcredit interest rates'*, CGAP brief, World Bank, Washington, DC, <https://openknowledge.worldbank.org/handle/10986/9510 License: CC BY 3.0 IGO>.

Lorentz (2020a) *Efficiency of solar powered water systems*, <https://www.lorentz.de/products-and-technology/technology/efficiency>.

Lorentz (2020b) *NPSH Calculation*, <https://partnernet.lorentz.de/files/lorentz_psk2-cs_manual_en.pdf>.

MOAIWD (2012) *Standard operating procedures for aquifer pumping tests*, <https://www.rural-water-supply.net/_ressources/documents/default/1-807-4-1530191157.pdf>.

Moving Energy Initiative (2018) *A Summary of Technology-enabled Finance for Solar Systems in the Sahel: Burkina Faso*.

Mukherji, A., et al. (2017a) *Solar powered irrigation pumps in South Asia: Challenges, opportunities and the way forward*.

Mukherji, A., Chowdhury, D.R., Fishman, R., Lamichhane, N., Khadgi, V. and Bajracharya, S. (2017b) *'Sustainable financial solutions for the adoption of solar powered irrigation pumps in Nepal's Terai'*, ICIMOD, Kathmandu, <lib.icimod.org/record/32565>.

Performance at low irradiance: Technical information for Solibro SL2 CIGS thin-film module (downloaded from <https://solibro-solar.com in 2019>).

Schnepf, R. (2006) *'Agriculture-based renewable energy production'*, US Congressional Research Service, The Library of Congress, Washington, DC, <https://digital.library.unt.edu/ark:/67531/metacrs8685/m1/1/high_res_d/RL32712_2006Feb28.pdf>.

SABCS (2015) *Solar America Board for Codes and Standards*, <http://www.solarabcs.org/about/publications/reports/module-grounding/>.

Shah, T., Rajan, A., Prakash Rai, G., Verma, S. and Durga, N. (2018) 'Solar pumps and South Asia's energy-groundwater nexus: exploring implications and reimagining its future', *Environmental Research Letters*, 13(11), <https://iopscience.iop.org/article/10.1088/1748-9326/aae53f/meta#acknowledgements>.

Skinner, B. (2001) *Chlorinating Small Water Supplies: A Review of Gravity-powered and Water-powered Chlorinators*, London School of Hygiene and Tropical Medicine and Water, Engineering and Development Centre, London and Loughborough, <https://www.lboro.ac.uk/orgs/well/resources/well-studies/full-reports-pdf/task0511.pdf>.

UNICEF (2016) *'Scaling up solar powered water supply systems: review of experiences'*, UNICEF, New York, <www.unicef.org/wash/files/UNICEF_Solar_Powered_Water_System_Assessment.pdf>.

World Bank (2010) *Photovoltaics for Community Service Facilities: Guidance for Sustainability,* World Bank, Washington, DC, <http://documents.worldbank.org/curated/en/837791468332067596/Photovoltaics-for-community-service-facilities-guidance-for-sustainability>.

World Bank (2017) *'Solar water pumping 101 – How to protect solar panels from theft'*, World Bank, Washington, DC <https://www.worldbank.org/en/news/video/2017/03/21/solar-water-pumping-101-how-to-protect-solar-panels-from-theft>.

World Bank (2018a) Employment in agriculture (% of total employment) (modeled ILO estimate). <https://data.worldbank.org/indicator/SL.AGR.EMPL.ZS?most_recent_value_desc=false>.

World Bank (2018b) *New country classifications by income level: 2018-2019 by World Bank Data team*, <https://blogs.worldbank.org/opendata/new-country-classifications-income-level-2018-2019>.

World Bank (2018c) *Solar pumping: the basics*, <http://documents.worldbank.org/curated/en/880931517231654485/Solar-pumping-the-basics>.

World Bank (2019) *'Real interest rate (%)'*, <http://data.worldbank.org/indicator/FR.INR.RINR?year_high_desc=false>.

WSTF (2017) *Service Delivery Model Toolkit for Sustainable Water Supply Service*, Water Sector Trust Fund, Nairobi, <https://waterfund.go.ke/publications?download=88:service-delivery-models-toolkit-v2>.

Websites

Databases

POWER Data Access Viewer <https://power.larc.nasa.gov/data-access-viewer/>.
Solargis solar resource maps <https://solargis.com/maps-and-gis-data/overview>.
<https://solargis.com/maps-and-gis-data/download/kenya>. (Annex B)

Manufacturers of SPWS equipment (Chapter 3)

ABB solar drives: https://new.abb.com/drives/low-voltage-ac/machinery/ABB-solar-pump-drives
Franklin solar pumps and drives: https://solar.franklin-electric.com/products/high-efficiency/6-inch-high-efficiency-solar-system/
Fuji – solar drives: https://www.fujielectric-europe.com/en/drives_automation/products/solutions/frenic_ace_for_solar_pumping
Grundfos company: www.grundfos.com
Lorentz pumps and controllers: www.lorentz.com
Solar service provider: https://www.davisandshirtliff.com
Solar service provider: https://solargentechnologies.com
Well pumps company: https://wellpumps.eu/en/homepage

PV module datasheets (accessed June 2019)

s-Si: Trina Solar Tallmax DE15M(II) 415 W (www.trinasolar.com)
p-Si: Trina Solar Honey PE06H 300 W(www.trinasolar.com)
a-Si/μc-Si: Kaneka Hybrid U-EA type 120 W (http://www.kaneka-solar.com/product/thin-film/)
CdTe: First Solar Series 6 FS-6445 (http://www.firstsolar.com/Modules/Series-6)
CIS: Solar Frontier SFK185-S 185 W (http://www.solar-frontier.com/eng/)
CIGS: Solibro SL2 Generation 2.3 150 W (https://solibro-solar.com)
HIT: Panasonic VBHN340SJ53 340 W (https://eu-solar.panasonic.net/en/)